Dictionary of
Pharmaceutical
Dosage Forms

Dictionary of Pharmaceutical Dosage Forms

Jeffrey T. Solate

CRC Press
Taylor & Francis Group
Boca Raton   London   New York

CRC Press is an imprint of the
Taylor & Francis Group, an **informa** business

First edition published 2021
by CRC Press
6000 Broken Sound Parkway NW, Suite 300, Boca Raton, FL 33487-2742

and by CRC Press
2 Park Square, Milton Park, Abingdon, Oxon, OX14 4RN

Library of Congress Cataloging-in-Publication Data
Names: Solate, Jeffrey T., author.
Title: Dictionary of pharmaceutical dosage forms / Jeffrey T. Solate.
Description: Boca Raton : CRC Press, 2020. | Includes bibliographical references and index.
Identifiers: LCCN 2019058620 (print) | LCCN 2019058621 (ebook) | ISBN 9781138065796 (paperback) | ISBN 9781315159447 (ebook)
Subjects: LCSH: Drugs--Dosage forms--Dictionaries.
Classification: LCC RS200 .S65 2020 (print) | LCC RS200 (ebook) | DDC 615.1/9003--dc23
LC record available at https://lccn.loc.gov/2019058620
LC ebook record available at https://lccn.loc.gov/2019058621

ISBN: 978-1-138-06579-6 (pbk)
ISBN: 978-1-315-15944-7 (ebk)

Typeset in Palatino
by Lumina Datamatics Limited

To my wonderful wife whose matchless support allowed me time to prepare this book. I'm also blessed by the Lord to share this new book with my daughter, Bridgette, my son, Theodore, my grandchildren, Taijha, Talia, and Nayali, and my son-in-law, Josh, who consistently demonstrated a special interest in family healthcare.

Contents

Introduction

The study of pharmaceutical dosage forms is not a new field but has many connections to several biological and medical sciences, including physiology, biochemistry, pharmacology, pharmacotherapy, clinical therapeutics, pharmacodynamics, pharmacokinetics, and pharmacognosy.

This first edition of *Dictionary of Pharmaceutical Dosage Forms* is a collection of terms and definitions prepared to assist healthcare practitioners and students and interns in Healthcare Services as a companion or reference resource when reading notes and completing routine care.

The dictionary can also be used as reference material by hospital and medical-center personnel, consultants, medical and nursing instructors, and pharmacology and pharmaceutical science students.

This dictionary also attempts to classify and organize vaccines and vaccination terms from various sources in an alphabetical listing.

The reader will find it a perfect companion to other technical and industry materials.

Jeffrey T. Solate
Pharmaceutical and Medical Device Lead Auditor,
Former, ASQ Food, Drug and Cosmetic Division, and
Drug Advisory Committee Public Member, Ontario College of Pharmacists

Formulation: Pills
Route of administration: Oral

Description of pills: A pill is a small, round, solid pharmacological oral dosage form in use before the advent of tablets and capsules. Pills were made by mixing the active ingredients with an excipient, such as glucose syrup in a mortar and pestle to form a paste, then rolling the mass into a long cylindrical shape (called a "pipe"), and dividing it into equal portions, which were then rolled into balls, and often coated with sugar to make them more palatable. In colloquial usage, tablets, capsules, and caplets are still often referred to as "pills" collectively.

Fluoxetine (Prozac)

Fluoxetine	How supplied	• 10 mg Pulvule is an opaque green cap and opaque green body. • 20 mg Pulvule is an opaque green cap and opaque yellow body. • 40 mg Pulvule is an opaque green cap and opaque orange body. • 90 mg Prozac Weekly™ Capsule is an opaque green cap and clear body containing discretely visible white pellets through the clear body of the capsule.
	Dosage	**Adult:** In controlled trials used to support the efficacy of Fluoxetine, patients were administered morning doses ranging from 20 to 80 mg/day. Studies comparing Fluoxetine 20, 40, and 60 mg/day to placebo indicate that 20 mg/day is sufficient to obtain a satisfactory response in Major Depressive Disorder in most cases. Consequently, a dose of 20 mg/day, administered in the morning, is recommended as the initial dose.

		A dose increase may be considered after several weeks if insufficient clinical improvement is observed. Doses above 20 mg/day may be administered on a once-a-day (morning) or BID schedule (i.e., morning and noon) and should not exceed a maximum dose of 80 mg/day. **Pediatric (children and adolescents):** In the short term (8–9 weeks), controlled clinical trials of Fluoxetine supporting its effectiveness in the treatment of Major Depressive Disorder, patients were administered Fluoxetine doses of 10–20 mg/day [see Clinical Studies]. Treatment should be initiated with a dose of 10 or 20 mg/day. After 1 week at 10 mg/day, the dose should be increased to 20 mg/day. However, due to higher plasma levels in lower-weight children, the starting and target dose in this group may be 10 mg/day. A dose increase to 20 mg/day

may be considered after several weeks if insufficient clinical improvement is observed.

Daily Dosing: Systematic evaluation of Prozac in adult patients has shown that its efficacy in Major Depressive Disorder is maintained for periods of up to 38 weeks following 12 weeks of open-label acute treatment (50 weeks total) at a dose of 20 mg/day.

Weekly Dosing: Systematic evaluation of Prozac® Weekly™ in adult patients has shown that its efficacy in Major Depressive Disorder is maintained for periods of up to 25 weeks with once-weekly dosing following 13 weeks of open-label treatment with Prozac 20 mg once daily. However, therapeutic equivalence of Prozac Weekly given on a once-weekly basis with Prozac 20 mg given daily for delaying time to relapse has not been established.

		Weekly dosing with Prozac Weekly capsules is recommended to be initiated 7 days after the last daily dose of Prozac 20 mg.
	Use	Fluoxetine is indicated for the acute and maintenance treatment of Major Depressive Disorder, obsessions, compulsions, binge-eating and vomiting behaviors, Panic Disorder.
	Patient group	Adults and children for over the age of 7.
	Side effects	Headache; asthenia; nausea; diarrhea; anorexia; insomnia; dizziness; dry mouth; increased sweating; loss of appetite; nervousness; stomach upset; trouble sleeping; weakness; severe allergic reactions (rash; hives; itching; difficulty breathing; tightness in the chest; swelling of the mouth, face, lips, or tongue); bizarre behavior; black or bloody stools; chest pain; confusion; exaggerated reflexes; excessive sweating; fast or irregular

| | | heartbeat; fever, chills, or sore throat; hallucinations; increased urination; joint or wrist aches or pain; loss of coordination; new or worsening agitation, panic attacks, aggressiveness, impulsiveness, irritability, hostility, exaggerated feeling of well-being, restlessness, or inability to sit still; persistent or severe ringing in the ears; persistent, painful erection; red, swollen, blistered, or peeling skin; seizures; severe or persistent anxiety or trouble sleeping; significant weight loss; stomach pain; suicidal thoughts or attempts; tremor; unusual bruising or bleeding; unusual hoarseness; unusual or severe mental or mood changes; unusual swelling; vision changes; worsening of depression. |
| | Half life | 1-3 days (acute admin.)
4-6 days (chronic admin.) |

Route of administration: Oral

Birth Control Pills

Birth Control Pills	Dosage	Birth control pills generally come packaged in blister packs containing 21 or 28 tablets: After finishing the 21-day pack, there is a week without any pills when withdrawal bleeding occurs. This will be like a regular menstrual period but generally lighter. Then a new pack is started on the same day of the week as the start of the previous pack. In the 28-day pack, typically 21 pills are active pills containing hormones and the other pills are "reminder" pills without hormones and are called spacer pills. Some reminder pills may contain iron. By taking reminder pills, a woman takes a pill each day, which can help maintain her birth control regimen. It is possible to reduce the number of menstrual periods a woman has in a year or even to eliminate them all together.

Since 2003, brands of birth control pills have been available in 84-day regimens for what's known as extended-cycle oral contraception. Women who use this method take an active pill every day for 12 weeks. Then they may stop or take a placebo for 7 days, during which time they will have their period. Or they may continue to take an active pill for up to a year. Some brands of birth control pills sold for extended-cycle oral contraception conclude the regimen with seven active pills that contain a smaller dose of estrogen, which can have the effect of reducing the amount of bleeding that occurs during a period.

Birth control pills can differ not only in the number of active ingredients but also in the way ingredients are dosed:

Monophasic: birth control pills contain the same amount of ingredient in each active pill.

		Multiphasic: birth control pills contain varying levels of hormones through the month. They were designed to minimize side effects, such as breakthrough bleeding, which is bleeding that occurs between menstrual periods. Low-dose oral contraceptives contain less estrogen than other types of birth control pills. They contain 20 µg of estrogen, compared to 30–50 µg in other birth control pills.
	Use	They offer protection against pregnancy by blocking the union of sperm and egg, thereby preventing conception. Oral contraceptives do not protect against HIV or other sexually transmitted infections.
	Patient group	Women
	Side effects	In healthy women, oral contraceptives have few side effects. Nausea, breast tenderness, weight gain,

changes in mood, and breakthrough bleeding are the most common ones—and these usually diminish with continued use. Pregnancy is still possible. With proper use, though, that possibility is minimized.

The following symptoms may indicate serious—even life-threatening—side effects: Pain in the chest or abdomen, severe headache, blurry vision, pain and/or swelling in the legs or thighs.

A woman who develops any of those symptoms should seek immediate medical attention.

Though oral contraceptives are usually well-tolerated in healthy women, there can be serious complications associated with their use. Some of the serious conditions include:

Cardiovascular: Blood clot in the veins, arteries, or lungs; heart attack; hypertension; Central nervous system; stroke; blood clot in the eye

| | | **Gastrointestinal:** Benign and cancerous tumors of the liver; blood clot in the blood vessels that support the intestines; gallbladder disease |
| | | For women with previous health issues, birth control pills may be a poor choice for contraception. The biggest concern is the generation of blood clots, strokes, and heart attacks—especially in women who are older and who smoke. In fact, women who smoke and take birth control pills dramatically increase their risk of developing strokes and heart attacks. The risk increases with age and amount of cigarette use. Birth control pills that contain estrogen may worsen diabetes. Women who experience migraine headaches, particularly those over age 35 and those who experience migraines with visual symptoms, are also at increased risk of stroke when using oral contraceptives. |

		Other medical history items that would prohibit oral contraceptive use include: blood clots; liver disease; breast or uterine cancer; hypertension; severe or uncontrolled diabetes. There is some evidence that long-term use of birth control pills may increase the risk for cervical cancer. And while studies show a slightly higher risk for breast cancer in women who have used the pill, no conclusions have been reached. Oral contraception may increase liver cancer risk. Oral contraceptives have been shown in studies to reduce the risk of ovarian and endometrial cancers
	Half-life	Birth control pills have a short HL, meaning if you are supposed to have it at 8:00 a.m. but take it at 4:00 p.m., you could be unprotected for hours.

Formulation: Tablets
Route of administration: Oral

Description of tablets: A tablet comprises a mixture of active substances and excipients, usually in powder form, pressed or compacted into a solid. The excipients can include binders, glidants (flow aids), and lubricants to ensure efficient tableting; disintegrants to promote tablet breakup in the digestive tract; sweeteners or flavors to enhance taste; and pigments to make the tablets visually attractive. A polymer coating is often applied to make the tablet smoother and easier to swallow, to control the release rate of the active ingredient, to make it more resistant to the environment (extending its shelf life), or to enhance the tablet's appearance.

The compressed tablet is the most popular dosage form in use today. About two-thirds of all prescriptions are dispensed as solid dosage forms, and half of these are compressed tablets. A tablet can be formulated to deliver an accurate dosage to a specific site. Medicinal tablets were originally made in the shape of a disk of whatever color their components determined but are now made in many shapes and colors to distinguish between different medicines that they take. Tablets are often stamped with symbols, letters, and numbers, which enable them to be identified. Sizes of tablets to be swallowed range from a few millimeters to about a centimeter.

Benadryl

Benadryl	Dosage	Ultratabs® tablets: Box of 24s, 48s, and 100s Kapseals® capsules: Box of 24s and 48s
	Use	Runny nose, sneezing, itchy and watery eyes, itchy throat, fever. Benadryl® Allergy contains the histamine-blocker diphenhydramine.

Patient group	Adults and children 12 years of age and over: 25–50 mg (1–2 capsules). Children 6 to under 12 years of age: 12.5–25 mg (1 capsule). Children under 6 years of age consult a doctor	
Side effects	Marked drowsiness and excitability especially in children Dry mouth, throat, and nose; Thickening of mucus in nose or throat. Allergic reactions (rash, hives, itching) Difficulty breathing Tightness in the chest Swelling of the mouth, face, lips, or tongue Convulsions Fast heartbeat or pounding in the chest Decreased alertness Hallucinations Tremor and wheezing	
Half-life	Generally 3.4–9.2 hours	

Formulation: Tablets
Route of administration: Oral

Codeine

Codeine	Dosage	15 mg white scored tablets Unit dose, 25 tablets per card, 4 cards per shipper 30 mg white scored tablets Unit dose, 25 tablets per card, 4 cards per shipper and bottles of 100 tablets 60 mg white scored tablets 25 tablets per card, 4 cards per shipper, and bottles of 100 tablets.
	Use	Cough Diarrhea Mild to moderate pain Irritable bowel syndrome
	Patient group	Adults
	Side effects	Analgesia, euphoria, itching, nausea, vomiting, drowsiness, dry mouth, miosis, orthostatic hypotension, urinary retention, depression, and constipation. Another side effect commonly noticed is the lack of sexual drive and increased complications in erectile dysfunction.

		Some people may also have an allergic reaction to codeine, such as the swelling of skin and rashes.
	Half-life	Plasma 2.9 hours

Formulation: Capsules
Route of administration: Oral

Description of capsules: Encapsulation refers to a range of techniques used to enclose medicines in a relatively stable shell known as a capsule, allowing them to, for example, be taken orally or be used as suppositories. The two main types of capsules are hard-shelled capsules, which are normally used for dry, powdered ingredients, and soft-shelled capsules, primarily used for oils and for active ingredients that are dissolved or suspended in oil. Both of these classes of capsules are made from gelatine and plant-based gelling substances like carrageenans and modified forms of starch and cellulose.

Since their inception, capsules have been viewed by consumers as the most-efficient method of taking medication. For this reason, producers of drugs wanting to emphasize the strength of their product developed the "caplet" or "capsule-shaped tablet" in order to tie this positive association to more efficiently produced tablet pills.

Soft-gel encapsulation

In 1834, Mothes and Dublanc were granted a patent for a method to produce a single-piece gelatin capsule that was sealed with a drop of gelatin solution. They used individual iron moulds for their process, filling the capsules individually with a medicine dropper. Later on, methods were developed that used sets of plates with pockets to form the capsules. Although some companies still use this method, the equipment is not produced commercially any more. All modern soft-gel encapsulations use variations of a process developed by R. P. Scherer in 1933. His innovation was to use a rotary die to produce the capsules, with the filling taking place by blow molding. This method reduced wastage and was the first process to yield capsules with highly repeatable dosage.

Soft gels can be an effective delivery system for oral drugs, especially poorly soluble drugs. This is because the fill can contain liquid ingredients that help increase solubility or permeability of the drug across the membranes in the body. Liquid ingredients are difficult to include in any other solid dosage form such as a tablet. Soft gels are also highly suited to potent drugs (e.g., where the dose is <100 µg), where the highly reproducible filling process helps ensure each soft gel has the same drug content, and because the operators are not exposed to any drug dust during the manufacturing process.

Two-part gel capsules

James Murdock patented the two-part telescoping gelatin capsule in London in 1847. The capsules are made in two parts by dipping metal rods in molten starch or cellulose solution. The capsules are supplied as closed units to the pharmaceutical manufacturer. Before use, the two halves are separated, the capsule is filled with powder, and the other half of the capsule is pressed on. The advantage of inserting a slug of compressed powder is that control of weight variation is better, but the machinery involved is more complex.

Doxycycline is a member of the tetracycline antibiotics group and is commonly used to treat a variety of infections. It works by slowing the growth of bacteria; slowing bacteria's growth allows the body's immune system to destroy the bacteria.

Doxycycline	Dosage	**Adults:** The usual dose of oral doxycycline is 200 mg on the first day of treatment (administered 100 mg every 12 hours or 50 mg every 6 hours) followed by a maintenance dose of 100 mg/day. The maintenance dose may be administered as a single dose or as 50 mg every 12 hours. In the management of more severe infections (particularly chronic infections of the urinary tract), 100 mg every 12 hours is recommended.

For pediatric patients above 8 years of age: The recommended dosage schedule for pediatric patients weighing 100 pounds or less is 2 mg/lb of body weight divided into two doses on the first day of treatment, followed by 1 mg/lb of body weight given as a single daily dose or divided into two doses, on subsequent days. For more severe infections, up to 2 mg/lb of body weight may be used. For pediatric patients over 100 pounds, the usual adult dose should be used.

Uncomplicated gonococcal infections in adults (except anorectal infections in men): 100 mg, by mouth, twice a day for 7 days. As an alternate single visit dose, administer 300 mg stat followed in 1 hour by a second 300 mg dose.

Acute epididymo-orchitis caused by *N. gonorrhoeae*: 100 mg, by mouth, twice a day for at least 10 days.

Primary and secondary syphilis: 300 mg a day in divided doses for at least 10 days.

| | | Uncomplicated urethral, endocervical, or rectal infection in adults caused by *Chlamydia trachomatis*: 100 mg, by mouth, twice a day for at least 7 days. **Nongonococcal urethritis caused by *C. trachomatis* and *Ureaplasma urealyticum***: 100 mg, by mouth, twice a day for at least 7 days. **Acute epididymo-orchitis caused by *C. trachomatis***: 100 mg, by mouth, twice a day for at least 10 days.

Inhalational anthrax (post-exposure). **Adults:** 100 mg of doxycycline, by mouth, twice a day for 60 days. **Children:** weighing less than 100 pounds (45 kg): 1 mg/lb (2.2 mg/kg) of body weight, by mouth, twice a day for 60 days. Children weighing 100 pounds or more should receive the adult dose. When used in streptococcal infections, therapy should be continued for 10 days. |
| | Use | To reduce the development of drug-resistant bacteria and maintain the effectiveness of doxycycline capsules and other antibacterial drugs, doxycycline capsules should |

be used only to treat or prevent infections that are proven or strongly suspected to be caused by susceptible bacteria. When culture and susceptibility information are available, they should be considered in selecting or modifying antibacterial therapy. In the absence of such data, local epidemiology and susceptibility patterns may contribute to the empiric selection of therapy. Doxycycline is also indicated for the treatment of infections caused by the following **gram-negative microorganisms**:

- Chancroid caused by *Haemophilus ducreyi.*
- Plague due to *Yersinia pestis* (formerly *Pasteurella pestis*)
- Tularemia due to *Francisella tularensis* (formerly *Pasteurella tularensis*)
- Cholera caused by *Vibrio cholerae* (formerly *Vibrio comma*)

Doxycycline is indicated for treatment of infections caused by the following **gram-negative microorganisms**, when

| | | bacteriologic testing indicates appropriate susceptibility to the drug:
 • *Escherichia coli*
 • *Enterobacter aerogenes* (formerly *Aerobacter aerogenes*)
 • *Shigella* species
 • *Acinetobacter* species (formerly *Mima* species and *Herellea* species)

 Doxycycline is not the drug of choice in the treatment of any type of staphylococcal infections. When penicillin is contraindicated, doxycycline is an alternative drug in the treatment of the following infections:
 • Uncomplicated gonorrhea caused by *Neisseria gonorrhoeae*.
 • Syphilis caused by *Treponema pallidum*.
 • Yaws caused by *Treponema pertenue*.
 • Listeriosis due to *Listeria monocytogenes*.
 • Vincent's infection caused by *Fusobacterium fusiforme*. |
| | Patient group | Adults and pediatric patients above 8 years of age |

Side effects	Due to oral doxycycline's virtually complete absorption, side effects to the lower bowel, particularly diarrhea, have been infrequent. Gastrointestinal: Anorexia, nausea, vomiting, diarrhea, glossitis, dysphagia, enterocolitis, and inflammatory lesions (with monilial overgrowth) in the anogenital region. Rare instances of esophagitis and esophageal ulcerations have been reported in patients receiving capsule and tablet forms of drugs in the tetracycline class. Most of these patients took medications immediately before going to bed. Skin: Maculopapular and erythematous rashes. Exfoliative dermatitis has been reported but is uncommon. Renal toxicity: Rise in BUN has been reported and is apparently dose related. Hypersensitivity reactions: Urticaria, angioneurotic edema, anaphylaxis, anaphylactoid purpura, pericarditis, and exacerbation of systemic lupus erythematosus.

		Blood: Hemolytic anemia, thrombocytopenia, neutropenia, and eosinophilia have been reported with tetracyclines. Other: Bulging fontanels in infants and intracranial hypertension in adults. When given over prolonged periods, tetracyclines have been reported to produce brown-black microscopic discoloration of the thyroid gland. No abnormalities of thyroid function are known to occur.
	Half-life	Approx. 24 hours.

Lithium Citrate is a chemical compound of lithium and citrate that is used as a mood stabilizer in psychiatric treatment of manic states and bipolar disorder.

Lithium Citrate	Dosage	**For acute manic episodes:** 1.8 g or 20–30 mg per kg of lithium carbonate per day in two to three divided doses. Some health-care providers begin therapy at 600–900 mg per day and gradually increase the dose.

		For bipolar disorder and other psychiatric conditions: The usual adult dose is 900 mg to 1.2 g per day in two to four divided doses. Approximately 24–32 mEq of lithium citrate solution, given in two to four divided doses daily, has also been used. Doses usually should not exceed 2.4 g of lithium carbonate or 65 mEq lithium citrate daily. **For children:** 15–60 mg per kg (0.4–1.6 mEq per kg) per day in divided doses has been used. Lithium may be given as a single daily dose, but is usually given in divided doses to lessen side effects. Stopping lithium therapy suddenly increases the chance that symptoms of bipolar disorder will return. The dose of lithium should be reduced gradually over at least 14 days. There is no recommended dietary allowance (RDA) for lithium. A provisional RDA of 1 mg per day for a 70 kg adult has been suggested.

Use	Bipolar Disorder (Manic-Depressive Disorder). Depression. Schizophrenia and related mental disorders. Impulsive aggressive behavior. Aggression associated with Attention-Deficit Hyperactivity Disorder (ADHD).	
Patient group	Adults and children	
Side effects	Weight gain; it increases the appetite and thirst, reduces the activity of thyroid hormone. Lithium is also a strong teratogen and if taken during a woman's pregnancy can cause her child to develop Ebstein's anomaly (cardiac defect). Lithium can cause nausea, diarrhea, dizziness, muscle weakness, fatigue, and a dazed feeling. These adverse effects often improve with continued use.	
Half-life	24 hours	

Formulation: Thin film
Route of administration: Oral

Description of film: Thin-film drug delivery is a process of delivering drugs to the systemic circulation via a thin film that dissolves when in contact with liquid, often referred to as a dissolving film or strip. Thin-film drug delivery has emerged as an advanced alternative to the traditional tablets, capsules, and liquids often associated with prescription and over-the-counter (OTC) medications. Similar in size, shape, and thickness to a postage stamp, thin-film strips are typically designed for oral administration, with the user placing the strip on or under the tongue or along the inside of the cheek. As the strip dissolves, the drug can enter the blood stream enterically, buccally, or sublingually.

The first commercial non-drug product to use thin films was the "Listerine PocketPacks" breath freshening strips. Since then, thin-film products for other breath fresheners, as well as a number of cold, cough, flu, and anti-snoring medications, have entered the marketplace. There are currently several projects in development that will deliver prescription drugs utilizing the thin-film dosage form.

Listerine PocketPacks

	Dosage	Read pack label and instruction for use
	Use	Kills germs that cause bad breath, reduce plaque, strengthen teeth to prevent cavities, prevent tartar buildup to keep teeth white, prevent gingivitis, and freshen breath for up to 12 hours.
Listerine PocketPacks	Patient group	Children and adults
	Side effects	Not applicable
	Half-life	Not applicable. Absorbed in the mouth

Cordran® Tape

Cordran® Tape (Flurandrenolide) Tape, USP	Drug description	Cordran® Tape (Flurandrenolide Tape, USP) is a transparent, inconspicuous, plastic surgical tape, impervious to moisture. It contains Cordran® (Flurandrenolide, USP), a potent corticosteroid for topical use. Flurandrenolide occurs as white to off-white, fluffy crystalline powder and is odorless. Flurandrenolide is practically insoluble in water and in ether. One gram dissolves in 72 mL of alcohol and in 10 mL of chloroform.
	How supplied	• 4 mcg/sq cm-small roll, 24 in × 3 in (60 × 7.5 cm) NDC 55515-014-24 • 4 mcg/sq cm-large roll, 80 in × 3 in (200 × 7.5 cm) NDC 55515-014-80

Dosage	Occlusive dressings may be used for the management of psoriasis or recalcitrant conditions. If an infection develops, the use of Cordran Tape and other occlusive dressings should be discontinued and appropriate antimicrobial therapy instituted.
Use	For relief of the inflammatory and pruritic manifestations of corticosteroid-responsive dermatoses, particularly dry, scaling localized lesions.
Patient group	Adults and children
Side effects	The following local adverse reactions are reported infrequently with topical corticosteroids but may occur more frequently with the use of occlusive dressings. These reactions are listed in an approximate decreasing order of occurrence:

| | | Burning, itching, irritation, dryness, folliculitis, hypertrichosis, acneform eruptions, hypopigmentation, perioral dermatitis, allergic contact dermatitis. The following may occur more frequently with occlusive dressings: Maceration of the skin; secondary infection, skin atrophy, striae, miliaria |
| | Half-life | Generally 0.5 hours |

Formulation: Syrup (liquid solution or suspension)
Route of administration: Oral

Description of syrup: Solution is a homogeneous mixture composed of two or more substances.

Celestone® (Betamethasone) Syrup

Celestone® (Betamethasone) Syrup	How supplied	Celestone Syrup, 0.6 mg per 5 mL, orange-red-colored liquid, bottle of 4 fl oz (118 mL)
	Dosage	The initial dosage may vary from 0.6 mg to 7.2 mg per day depending on the specific disease entity being treated. In the treatment of acute exacerbations of multiple sclerosis, daily doses of 30 mg of Betamethasone for a week followed by 12 mg every other day for 1 month are recommended. In pediatric patients, the initial dose of Betamethasone may vary depending on the specific disease entity being treated. The range of initial doses is 0.02 to 0.3 mg/kg/day in three or four divided doses (0.6–9 mg/m^2 bsa/day).

	Use	**Allergic states:** Control of severe or incapacitating allergic conditions intractable to adequate trials of conventional treatment in asthma, atopic dermatitis, contact dermatitis, drug hypersensitivity reactions, perennial or seasonal allergic rhinitis, serum sickness. **Dermatologic diseases:** Bullous dermatitis herpetiformis, exfoliative erythroderma, mycosis fungoides, pemphigus, severe erythema multiforme (Stevens-Johnson syndrome). **Endocrine disorders:** Congenital adrenal hyperplasia, hypercalcemia associated with cancer, nonsuppurative thyroiditis. Hydrocortisone or cortisone is the drug of choice in primary or secondary adrenocortical insufficiency. Synthetic analogs may be used in conjunction with mineralocorticoids where applicable; in infancy, mineralocorticoid supplementation is of particular importance. **Gastrointestinal diseases:** To tide the patient over a critical period of the disease in regional enteritis and ulcerative colitis.

		Hematologic disorders: Acquired (autoimmune) hemolytic anemia, Diamond-Blackfan anemia, idiopathic thrombocytopenic purpura in adults, pure red cell aplasia, selected cases of secondary thrombocytopenia. **Miscellaneous:** Trichinosis with neurologic or myocardial involvement, tuberculous meningitis with subarachnoid block or impending block when used with appropriate antituberculous chemotherapy.
		Neoplastic diseases: For the palliative management of leukemias and lymphomas. **Nervous system:** Acute exacerbations of multiple sclerosis; cerebral edema associated with primary or metastatic brain tumor, craniotomy, or head injury. **Ophthalmic diseases**: Sympathetic ophthalmia, temporal arteritis, uveitis and ocular inflammatory conditions unresponsive to topical corticosteroids.

| | | **Renal diseases**: To induce diuresis or remission of proteinuria in idiopathic nephrotic syndrome or that due to lupus erythematosus. **Respiratory diseases**: Berylliosis, fulminating or disseminated pulmonary tuberculosis when used concurrently with appropriate antituberculous chemotherapy, idiopathic eosinophilic pneumonias, symptomatic sarcoidosis. **Rheumatic disorders**: As adjunctive therapy for short-term administration (to tide the patient over an acute episode or exacerbation) in acute gouty arthritis; acute rheumatic carditis; ankylosing spondylitis; psoriatic arthritis; rheumatoid arthritis, including juvenile rheumatoid arthritis (selected cases may require low-dose maintenance therapy). For the treatment of dermatomyositis, polymyositis, and systemic lupus erythematosus. |
| | Patient group | Adults and children |

	Side effects	**Allergic reactions**: Anaphylactoid reaction, anaphylaxis, angioedema. **Cardiovascular:** Bradycardia, cardiac arrest, cardiac arrhythmias, cardiac enlargement, circulatory collapse, congestive heart failure, fat embolism, hypertension, hypertrophic cardiomyopathy in premature infants, myocardial rupture following recent myocardial infarction, pulmonary edema, syncope, tachycardia, thromboembolism, thrombophlebitis, vasculitis. **Dermatologic**: Acne, allergic dermatitis, dry scaly skin, ecchymoses and petechiae, edema, erythema, impaired wound healing, increased sweating, rash, striae, suppressed reactions to skin tests, thin fragile skin, thinning scalp hair, urticaria. **Endocrine:** Decreased carbohydrate and glucose tolerance, development of cushingoid state, glycosuria, hirsutism, hypertrichosis, increased requirements for insulin or oral hypoglycemic

		agents in diabetes, manifestations of latent diabetes mellitus, menstrual irregularities, secondary adrenocortical and pituitary unresponsiveness (particularly in times of stress, as in trauma, surgery, or illness), suppression of growth in pediatric patients. **Fluid and electrolyte disturbances**: Congestive heart failure in susceptible patients, fluid retention, hypokalemic alkalosis, potassium loss, sodium retention.
		Gastrointestinal: Abdominal distention, elevation in serum liver enzyme levels (usually reversible upon discontinuation), hepatomegaly, increased appetite, nausea, pancreatitis, peptic ulcer with possible perforation and hemorrhage, perforation of the small and large intestine (particularly in patients with inflammatory bowel disease), ulcerative esophagitis. **Metabolic:** Negative nitrogen balance due to protein catabolism. **Musculoskeletal:** Aseptic necrosis of femoral and humeral heads, loss

		of muscle mass, muscle weakness, osteoporosis, pathologic fracture of long bones, steroid myopathy, tendon rupture, vertebral compression fractures. **Neurologic/Psychiatric:** Convulsions, depression, emotional instability, euphoria, headache, increased intracranial pressure with papilledema (pseudotumor cerebri) usually following discontinuation of treatment, insomnia, mood swings, neuritis, neuropathy, paresthesia, personality changes, psychic disorders, vertigo. **Ophthalmic:** Exophthalmos, glaucoma, increased intraocular pressure, posterior subcapsular cataracts. **Other:** Abnormal fat deposits, decreased resistance to infection, hiccups, increased or decreased motility and number of spermatozoa, malaise, moon face, weight gain.
	Half-life	Elimination HL: 36–54 hours

Clemastine Fumarate Syrup

Clemastine Fumarate Syrup	How supplied	Clemastine Fumarate Syrup: clemastine 0.5 mg/5 mL (present as clemastine fumarate 0.67 mg/5 mL). A clear, colorless liquid with a citrus flavor, in 120 mL bottle.
	Dosage	Adults and children 12 years and over: **For symptoms of allergic rhinitis**: The starting dose is 2 tsp (1 mg Clemastine) twice daily. Dosage may be increased as required, but not to exceed 12 tsp daily (6 mg Clemastine). **For urticaria and angioedema**: The starting dose is 4 tsp (2 mg Clemastine) twice daily, not to exceed 12 tsp daily (6 mg Clemastine).
	Use	Clemastine Fumarate Syrup is indicated for the relief of symptoms associated with allergic rhinitis, such as sneezing, rhinorrhea, pruritus, and lacrimation. Clemastine Fumarate Syrup is indicated for use in pediatric populations (age 6 years through 12) and adults.

		It should be noted that Clemastine is indicated for the relief of mild uncomplicated allergic skin manifestations of urticaria and angioedema at the 2-mg dosage level only.
	Patient group	Pediatric, children, and adults
	Side effects	The most-frequent adverse reactions are italicized: **Nervous system:** Sedation, sleepiness, dizziness, disturbed coordination, fatigue, confusion, restlessness, excitation, nervousness, tremor, irritability, insomnia, euphoria, paresthesia, blurred vision, diplopia, vertigo, tinnitus, acute labyrinthitis, hysteria, neuritis, convulsions. **Gastrointestinal system:** Epigastric distress, anorexia, nausea, vomiting, diarrhea, constipation.

| | | **Respiratory system:** Thickening of bronchial secretions, tightness of chest and wheezing, nasal stuffiness. **Cardiovascular system:** Hypotension, headache, palpitations, tachycardia, extrasystoles. **Hematologic system**: Hemolytic anemia, thrombocytopenia, agranulocytosis. **Genitourinary system**: Urinary frequency, difficult urination, urinary retention, early menses. **General:** Urticaria; drug rash; anaphylactic shock; photosensitivity; excessive perspiration; chills; dryness of mouth, nose, and throat. |
| | Half-life | Elimination HL: 21.3 hours +/− 11.6 hours |

Midazolam Hydrochloride Syrup

Midazolam Hydrochloride Syrup	How supplied	Midazolam HCL syrup is supplied as a clear, red to purplish-red, mixed-fruit-flavored syrup containing Midazolam Hydrochloride equivalent to 2 mg of Midazolam per mL; each amber glass bottle of 118 mL is supplied with 1 press-in bottle adapter, four single-use, graduated, oral dispensers, and four tip caps; 10 × bottle of 2.5 mL is supplied with 10 single-use, graduated, oral dispensers and 10 tip caps.
	Dosage	Midazolam HCL syrup is indicated for use as a single dose (0.25–1 mg/kg with a maximum dose of 20 mg) for preprocedural sedation and anxiolysis in pediatric patients. Midazolam HCL syrup is not intended for chronic administration.
	Use	Midazolam HCL syrup is indicated for use in pediatric patients for sedation, anxiolysis, and amnesia prior to diagnostic, therapeutic, or endoscopic procedures or before induction of anesthesia.

		Midazolam HCL syrup is intended for use in monitored settings only and not for chronic or home use. Midazolam is associated with a high incidence of partial or complete impairment of recall for the next several hours.
	Patient group	Pediatric patients
	Side effects	The adverse events that have been reported in the literature with the oral administration of Midazolam (not necessarily Midazolam Hydrochloride syrup) are listed in the following. The incidence rate for these events was generally <1%. **Respiratory**: apnea, hypercarbia, desaturation, stridor. **Cardiovascular**: decreased systolic and diastolic blood pressure, increased heart rate. **Gastrointestinal**: nausea, vomiting, hiccoughs, gagging, salivation, drooling. **Central nervous system**: dysphoria, disinhibition, excitation, aggression, mood swings, hallucinations, adverse behavior, agitation, dizziness,

		confusion, ataxia, vertigo, dysarthria. **Special senses**: diplopia, strabismus, loss of balance, blurred vision.
	Half-life	2.2–6.8 hours

Formulation: Powder, liquid, or solid
  crystals

Route of administration: Oral

Description:

Examples

	Dosage	Read label and Instructions for use
	Use	Read label and Instructions for use
	Patient group	All age groups.
	Side effects	Not applicable
	Half-life	Not applicable. Absorbed from the skin

Formulation: Natural or herbal plant, seed, or food

Route of administration: Oral

Description: Herbalism is a traditional medicinal or folk medicine practice based on the use of plants and plant extracts. Herbalism is also known as botanical medicine, medical herbalism, herbal medicine, herbology, and phytotherapy. The scope of herbal medicine is sometimes extended to include fungal and bee products, as well as minerals, shells, and certain animal parts.

Rosemary

Rosemary	Dosage	Not indicated
	Use	Rosemary shields the brain from free radicals, lowering the risk of strokes and neurodegenerative diseases like Alzheimer's and Lou Gehrig's.
	Patient group	All ages
	Side effects	Few instances of allergic skin reactions. Persons with iron deficiency anemia should not use it internally.
	Half-life	Not applicable

Formulation: Aerosol
Route of administration: Inhalational

Description of aerosol: Aerosol spray is a type of dispensing system that creates an aerosol mist of liquid particles. This is used with a can or bottle that contains a liquid under pressure. When the container's valve is opened, the liquid is forced out of a small hole and emerges as an aerosol or mist.

Flovent (Fluticasone Propionate)

Flovent (Fluticasone Propionate)	How supplied	**Flovent 44 mcg Inhalation Aerosol** is supplied in 7.9-g canisters containing 60 metered inhalations in institutional pack boxes of one (NDC 0173-0497-00) and in 13-g canisters containing 120 metered inhalations in boxes of one (NDC 0173-0491-00). Each canister is supplied with a dark orange oral actuator with a peach strapcap and patient instructions. Each actuation of the inhaler delivers 44 mcg of Fluticasone Propionate from the actuator.

Flovent 110 mcg Inhalation Aerosol is supplied in 7.9-g canisters containing 60 metered inhalations in institutional pack boxes of one (NDC 0173-0498-00) and in 13-g canisters containing 120 metered inhalations in boxes of one (NDC 0173-0494-00). Each canister is supplied with a dark orange oral actuator with a peach strapcap and patient instructions. Each actuation of the inhaler delivers 110 mcg of Fluticasone Propionate from the actuator.

Flovent 220 mcg Inhalation Aerosol is supplied in 7.9-g canisters containing 60 metered inhalations in institutional pack boxes of one (NDC 0173-0499-00) and in 13-g canisters containing 120 metered inhalations in boxes of one (NDC 0173-0495-00). Each canister is supplied with a dark orange oral actuator with a peach strapcap and patient instructions. Each actuation of the inhaler delivers 220 mcg of fluticasone propionate from the actuator.

	Dosage	**Bronchodilators:** Recommended starting: 88 mcg twice daily Highest recommended dosage: 440 mcg twice daily **Inhaled corticosteroids**: Recommended starting: 88–220 mcg twice daily Highest recommended dosage: 440 mcg twice daily **Oral corticosteroids**: Recommended starting: 880 mcg twice daily Highest recommended dosage: 880 mcg twice daily In all patients, it is desirable to titrate to the lowest effective dosage once asthma stability is achieved.
	Use	Flovent is used for preventing or reducing the frequency and seriousness of bronchial asthma attacks. Flovent does not help during an acute asthma attack. It is also indicated for patients requiring oral corticosteroid therapy for asthma. Many of these patients may be able to reduce or eliminate their requirement for oral corticosteroids over time.

		Flovent is a corticosteroid. It works by reducing inflammatory reactions in the airways in response to allergens and irritants in the air.
	Patient group	12 years of age and older
	Side effects	Aches and pains; asthma symptoms; bronchitis; changes in or loss of senses of smell and taste; cough; diarrhea; dizziness; fever; flu-like symptoms (headache, tiredness, muscle aches, fever); infection of the nose and throat; nasal irritation or burning; nausea; nosebleed; runny nose; sore throat; stomach pain; vomiting. Allergic reactions (rash; hives; itching; difficulty breathing; tightness in the chest; swelling of the mouth, face, lips, or tongue); cataracts; growth suppression in children; hoarseness; increased pressure in the eye; infection or pain in the nose or throat; nasal septum perforation; repeated nosebleeds; voice changes
	Half-life	Generally 8–15 hours

Formulation: Inhaler
Route of administration: Inhalational

Description of inhaler: An inhaler or puffer is a medical device used for delivering medication into the body via the lungs. It is mainly used in the treatment of asthma and Chronic Obstructive Pulmonary Disease

Advair Diskus
Entocort (Budesonide)

Advair Diskus	How supplied	Disposable purple device with 60 blisters containing a combination of Fluticasone Propionate (100, 250, or 500 mcg) and Salmeterol (50 mcg) as an oral inhalation powder formulation. An institutional pack containing 14 blisters is also available. **Advair Diskus 100/50** is supplied as a disposable purple device containing 60 blisters. The Diskus inhalation device is packaged within a purple, plastic-coated, moisture-protective foil pouch (NDC 0173-0695-00). Advair Diskus 100/50 is also supplied in an institutional pack of one disposable purple device

containing 14 blisters. The Diskus inhalation device is packaged within a purple, plastic-coated, moisture-protective foil pouch (NDC 0173-0695-04).

Advair Diskus 250/50 is supplied as a disposable purple device containing 60 blisters. The Diskus inhalation device is packaged within a purple, plastic-coated, moisture-protective foil pouch (NDC 0173-0696-00). Advair Diskus 250/50 is also supplied in an institutional pack of one disposable purple device containing 14 blisters. The Diskus inhalation device is packaged within a purple, plastic-coated, moisture-protective foil pouch (NDC 0173-0696-04).

Advair Diskus 500/50 is supplied as a disposable purple device containing 60 blisters. The Diskus inhalation device is packaged within a purple, plastic-coated, moisture-protective foil pouch (NDC 0173-0697-00). Advair Diskus 500/50 is also supplied in an institutional pack of one disposable purple device

		containing 14 blisters. The Diskus inhalation device is packaged within a purple, plastic-coated, moisture-protective foil pouch (NDC 0173-0697-04).
	Dosage	Advair Diskus should be administered twice daily every day by the orally inhaled route only. After inhalation, the patient should rinse the mouth with water without swallowing. More frequent administration or a higher number of inhalations (more than one inhalation twice daily) of the prescribed strength of Advair Diskus is not recommended because some patients are more likely to experience adverse effects with higher doses of Salmeterol. Patients using Advair Diskus should not use additional long-acting beta2-agonists for any reason.
		For asthma treatment: If asthma symptoms arise in the period between doses, an inhaled, short-acting beta2-agonist should be taken for immediate relief.

Adult and adolescent patients aged 12 years and older: the dosage is one inhalation twice daily (morning and evening, approximately 12 hours apart).

The recommended starting dosages for Advair Diskus for patients aged 12 years and older are based upon patients' asthma severity. For patients not currently on inhaled corticosteroids whose disease severity clearly warrants initiation of treatment with two maintenance therapies, or patients inadequately controlled on an inhaled corticosteroid, the recommended starting dosage is Advair Diskus 100/50 or 250/50 twice daily.

The maximum recommended dosage is Advair Diskus 500/50 twice daily.

For all patients, it is desirable to titrate to the lowest effective strength after adequate asthma stability is achieved.

		Pediatric patients aged 4–11 years: For patients with asthma aged 4–11 years who are symptomatic on an inhaled corticosteroid, the dosage is one inhalation of Advair Diskus 100/50 twice daily (morning and evening, approximately 12 hours apart). **Chronic Obstructive Pulmonary Disease** The recommended dosage for patients with COPD is one inhalation of Advair Diskus 250/50 twice daily (morning and evening, approximately 12 hours apart). If shortness of breath occurs in the period between doses, an inhaled, short-acting beta2-agonist should be taken for immediate relief.
	Use	Treatment of asthma
	Patient group	Children and adults with asthma Important limitations of use: • Advair Diskus is NOT indicated for the relief of acute bronchospasm.

		• Advair Diskus is NOT indicated in patients whose asthma can be successfully managed by inhaled corticosteroids along with occasional use of inhaled, short-acting beta2-agonists.
	Side effects	*Candida albicans* infection Pneumonia in patients with COPD Immunosuppression Hypercorticism and adrenal suppression Growth effects Glaucoma and cataracts **Additional adverse reactions:** Other adverse reactions not previously listed, whether considered drug-related or not by the investigators, that were reported more frequently by patients with asthma treated with Advair Diskus compared with patients treated with placebo include the following: lymphatic signs and symptoms; muscle injuries; fractures; wounds and lacerations; contusions and hematomas; ear signs and

| | | symptoms; nasal signs and symptoms; nasal sinus disorders; keratitis and conjunctivitis; dental discomfort and pain; gastrointestinal signs and symptoms; oral ulcerations; oral discomfort and pain; lower respiratory signs and symptoms; pneumonia; muscle stiffness, tightness, and rigidity; bone and cartilage disorders; sleep disorders; compressed nerve syndromes; viral infections; pain; chest symptoms; fluid retention; bacterial infections; unusual taste; viral skin infections; skin flakiness and acquired ichthyosis; disorders of sweat and sebum. |
| | Half-life | Average 5.33–7.65 hours |

Entocort (Budesonide)

<table>
<tr><td rowspan="6">Entocort (Budesonide)</td><td>Dosage</td><td>EC 3 mg</td></tr>
<tr><td>Use</td><td>Entocort inhalation is used to prevent asthma attacks. It will not treat an asthma attack that has already begun.
Entocort is a steroid. It prevents the release of substances in the body that cause inflammation.</td></tr>
<tr><td>Patient group</td><td>8 years and older</td></tr>
<tr><td>Side effects</td><td>Back pain, changes in menstrual cycle, dizziness, gas, headache, indigestion, nausea, nervousness, pain, respiratory tract infection, stomach pain, tiredness, tremor, trouble sleeping, vomiting.</td></tr>
<tr><td>Half-life</td><td>Average 4 hours</td></tr>
</table>

Aerobid (Flunisolide)

Aerobid (Flunisolide)	How supplied	Aerobid (Flunisolide) Inhaler Systems are available in canisters of 100 metered inhalations. NDC 0456-0672-99 AEROBID NDC 0456-0670-99 AEROBID-M
	Dosage	**Adults:** The recommended starting dose is two inhalations twice daily, morning and evening, for a total daily dose of 1 mg. The maximum daily dose should not exceed four inhalations twice a day for a total daily dose of 2 mg. When the drug is used chronically at 2 mg/day, patients should be monitored periodically for effects on the hypothalamic-pituitary-adrenal (HPA) axis. **Pediatric patients:** For children and adolescents 6–15 years of age, two inhalations may be administered twice daily for a total daily dose of 1 mg. Higher doses have not been studied.

		Insufficient information is available to warrant use in children **under age 6.** With chronic use, pediatric patients should be monitored for growth as well as for effects on the HPA axis. Rinsing the mouth after inhalation is advised.
	Use	Flunisolide is used to help prevent asthma symptoms such as wheezing and shortness of breath. For those who must take corticosteroids by mouth to control their asthma, Flunisolide can be used to help decrease the dose of the medication that must be taken by mouth. This medication belongs to a class of drugs known as corticosteroids. It works to make breathing easier by reducing the swelling of the airways in the lungs and decreasing phlegm and other asthma symptoms. This medication must be taken regularly to be effective. It does not work immediately and should not be used to relieve sudden asthma attacks.

Patient group	Children 6 years and over and adults	
Side effects	Stomach upset, nausea, vomiting, headache, sore throat, stuffy nose, or unpleasant taste in the mouth may occur.	
	Dizziness, light-headedness, shakiness, nervousness, white/sore patches in the mouth or throat, heartburn, loss of appetite, increased appetite, constipation, gas, menstrual period changes (e.g., delayed/irregular/absent periods), stomach/abdominal pain, fast/pounding/irregular heartbeat, unusual weight gain, swelling of the ankles/feet, mental/mood changes (e.g., depression, mood swings, agitation), fever, persistent sore throat, frequent/painful urination, eye pain/discharge, earache, cough, vision changes.	

		A very serious allergic reaction to this drug is unlikely, but seek immediate medical attention if it occurs. Symptoms of a serious allergic reaction may include: rash, itching, swelling, severe dizziness, trouble breathing.
	Half-life	1.3–1.7 hours

Albuterol Sulfate

Albuterol Sulfate	How supplied	Albuterol Sulfate Inhalation Solution is supplied as a 3 mL, clear, colorless, sterile, preservative-free, aqueous solution in two different strengths, 0.021% (0.63 mg) and 0.042% (1.25 mg) of Albuterol (equivalent to 0.75 mg of Albuterol Sulfate or 1.5 mg of Albuterol Sulfate per 3 mL) in unit-dose low-density polyethylene (LDPE) vials. Each unit-dose LDPE vial is protected in a foil pouch, and each foil pouch contains five unit-dose LDPE vials. Each strength of Albuterol Sulfate inhalation solution is available in a shelf carton containing multiple foil pouches. Albuterol Sulfate Inhalation Solution 0.021% (0.63 mg/3 mL) (potency expressed as Albuterol) contains 0.75 mg Albuterol Sulfate per 3 mL in unit-dose vials and is available in the following packaging configuration.

		NDC 0591-3467-53: 5 foil pouches, each containing five vials, total 25 vials per carton Albuterol Sulfate Inhalation Solution, 0.042% (1.25 mg/3 mL) (potency expressed as Albuterol) contains 1.50 mg Albuterol Sulfate per 3 mL in unit-dose vials and is available in the following packaging configuration. NDC 0591-3468-53: five foil pouches, each containing five vials, total 25 vials per carton.
	Dosage	The usual starting dosage for patients 2–12 years of age is 1.25 or 0.63 mg of Albuterol Sulfate Inhalation Solution administered three or four times daily, as needed, by nebulization. More-frequent administration is not recommended.

Use	Albuterol Sulfate Inhalation Solution is indicated for the relief of bronchospasm in patients 2–12 years of age with asthma. It works in the airways by opening breathing passages and relaxing muscles.
Patient group	For patients 2–12 years of age with asthma.
Side effects	Nervousness, shaking (tremor), mouth/throat dryness or irritation, cough, dizziness, headache, trouble sleeping, or nausea may occur.
Half-life	Elimination generally 7.8 minutes (Inhalation Aerosol)

Formulation: Nebulizer
Route of administration: Inhalational

Description of nebulizer: A nebulizer is a device used to administer medication to people in the form of a mist inhaled into the lungs. It is commonly used in treating cystic fibrosis, asthma, and other respiratory diseases.

Intal Nebulizer Solution: It is an inhaled anti-inflammatory agent for the preventive management of asthma.

Intal Nebulizer Solution (Cromolyn Sodium Inhalation Solution)	How supplied	Intal Nebulizer Solution is a colorless solution supplied in a low-density polyethylene plastic unit dose ampule with 12 ampules per foil pouch. Each 2-mL ampule contains 20 mg Cromolyn Sodium, USP, in purified water. NDC 0585-0673-02 ....60 ampules × 2 mL NDC 0585-0673-03 ....120 ampules × 2 mL

	Dosage	The usual starting dosage is the contents of one ampule administered by nebulization four times a day at regular intervals. In patients whose symptoms are sufficiently frequent to require a continuous program of medication, Intal is given by inhalation on a regular daily basis. In patients who develop acute bronchoconstriction in response to exposure to exercise, toluene diisocyanate, environmental pollutants, etc., Intal should be given shortly before exposure to the precipitating factor.
	Use	INTAL is a prophylactic agent indicated in the management of patients with bronchial asthma. The effect of Intal is usually evident after several weeks of treatment, although some patients show an almost immediate response.

Patient group	2 years of age and over
Side effects	Cough Nasal congestion Nausea Sneezing Wheezing Other reactions have been reported in clinical trials; however, a causal relationship could not be established: drowsiness, nasal itching, nose bleed, nose burning, serum sickness, and stomachache.
Half-life	Generally 81 minutes

NebuPent (pentamidine isethionate), an antiprotozoal agent, is a nonpyrogenic lyophilized product. After reconstitution with Sterile Water for Injection, USP, NebuPent is administered by inhalation via the Respirgard®II nebulizer

NebuPent (Pentamidine Isethionate)	How supplied	NebuPent (pentamidine isethionate) 300-mg lyophilized product in single-dose vials, individually packaged.
	Dosage	The recommended adult dosage of NebuPent for the prevention of *Pneumocystis carinii* pneumonia is 300 mg **once every 4 weeks** administered via the Respirgard®II nebulizer.
	Use	Pentamidine isethionate inhalation (prophylaxis) is used to prevent a serious lung infection (Pneumocystis pneumonia—PCP) in patients with acquired immunodeficiency syndrome (AIDS). It is used when the usual medications cannot be used. Pentamidine belongs to a class of drugs known as antimicrobials. It works by killing the organism that causes the infection.
	Patient group	Patients with acquired immunodeficiency syndrome (AIDS)

	Side effects	Cough Decreased appetite Dizziness or light-headedness Fatigue Fever Non-specific serious infection Shortness of breath Wheezing
	Half-life	Generally 6.4–9.4 hours

Formulation: Natural herb
Route of administration: Inhalational

Description: *Lobelia* is a plant. The above-ground parts are used to make medicine.

Lobelia	Dosage	The appropriate dose of lobelia depends on several factors such as the user's age, health, and several other conditions. At this time, there is not enough scientific information to determine an appropriate range of doses for lobelia.
	Use	Lobelia is a powerful anti-spasmodic and is the most powerful herb for relaxing the bronchial tubes and airway during an attack. A quality lobelia tincture is a more powerful anti-spasmodic than any asthma drug. Lobelia contains chemicals, which might thin mucus (phlegm) to make it easier to cough up (expectorate) and help breathing, especially in people with asthma. There are several other herbs that will dilate and relax the

		airways, but lobelia is by far the most powerful. Higher doses of lobelia will induce vomiting from the lungs, which can be an intense experience, but will clear some of the sludge that has built up in the lungs. • Use by mouth for asthma, bronchitis, cough, and other conditions. • Use on the skin for muscle soreness, bruises, sprains, insect bites, poison ivy, ringworm, and other conditions.
	Patient group	Children and adults
	Side effects	Side effects include nausea, vomiting, diarrhea, cough, dizziness, and tremors. Overdose may cause many serious toxic effects including sweating, convulsions, fast heartbeat, very low blood pressure, collapse, coma, and possibly death.
	Half-life	Not applicable

Formulation: Natural herb
Route of administration: Oral, inhalation, and as cream or lotion.

Cayenne Pepper: (Capsicum) is an herb. The fruit of the capsicum plant is used to make medicine.

Cayenne Pepper (Capsicum)	Dosage	For most chronic conditions, you should be taking three to six doses per day. Cayenne is an herb that you cannot take too much of, and it should form a core part of any asthma herbal treatment. Once you get used to it, cayenne should be something that you are taking on a daily basis until you are well.
	Use	Cayenne pepper is the most powerful herb for increasing circulation. When taken internally, it increases body-wide circulation. It also blasts circulation into any part of the body that it comes into contact with. All diseases are related to blockages within the body, and asthma is no exception.

		If you want to eliminate blockage, you will have to start by increasing circulation. When used as a lotion or cream and applied to the skin, Capsicum can be effective for: • Arthritis pain when applied to the skin • Pain from shingles when applied to the skin • Nerve pain (neuropathy) in people with diabetes when applied to the skin • Back pain • Reducing painful tender points in people with fibromyalgia • Relieving symptoms of prurigo nodularis (a skin disease) When used nasally, it can relieve: • Cluster headache
	Patient group	Adults and children
	Side effects	Cayenne can literally drive your face red Capsicum extract-containing **lotion or cream** is safe for most

		adults when applied to the skin. Side effects can include skin irritation, burning, and itching. Capsicum can also be extremely irritating to the eyes, nose, and throat. Don't use capsicum on sensitive skin or around the eyes.
		Capsicum extract seems to be safe for most adults when taken by **mouth**. Side effects can include stomach irritation and upset, sweating, flushing, and runny nose. Don't take capsicum by mouth in large doses or for long periods of time. In rare cases, this can lead to more serious side effects, such as liver or kidney damage. Capsicum extract seems to be safe when used **nasally**. No serious side effects have been reported, but application in the nose can be very painful. Nasal application can cause burning pain, sneezing, watery eyes, and runny nose.
	Half-life	Not applicable

Formulation: Freebase powder
Route of administration: Oral administration and inhalational

Description:

Viokase

	How supplied	**Powder**: Tan powder in bottles of 8 oz (227 g) (NDC 58914-115-08) **8 Tablets**: Tan, round, compressed tablets engraved Viokase on one side and 9111 on the other side in bottles of 100 (NDC 58914-111-10) and 500 (NDC 58914-111-50). **16 Tablets**: Tan, oval, biconvex tablets engraved V16 on one side and 9116 on the other side in bottles of 100 (NDC 58914-116-10).
Viokase	Dosage	**Powder**: Dosage for patients with cystic fibrosis: 1/4 teaspoonful (0.7 g) with meals.

		Tablets: Dosage range for patients with cystic fibrosis or chronic pancreatitis is from 8,000 to 32,000 Lipase USP Units taken with meals, i.e., one to four Viokase 8 tablets or one to two Viokase® 16 tablets with meals or as directed by a physician. In patients with pancreatectomy or obstruction of pancreatic ducts: one to two Viokase® 8 tablets or one Viokase® 16 tablet taken at 2-hour intervals or as directed by a physician
	Use	Viokase® (Pancrelipase, USP) is indicated in the treatment of exocrine pancreatic insufficiency as associated with but not limited to cystic fibrosis, chronic pancreatitis, pancreatectomy, or obstruction of the pancreas ducts.

	Patient group	Adult
	Side effects	The dust or finely powdered pancreatic enzyme concentrate is irritating to the nasal mucosa and the respiratory tract. It has been documented that inhalation of the airborne powder can precipitate an asthma attack. The literature also contains several references to asthma due to inhalation in patients sensitized to pancreatic enzyme concentrates. Extremely high doses of exogenous pancreatic enzymes have been associated with hyperuricemia and hyperuricosuria. Overdosage of pancreatic enzyme concentrate may cause diarrhea or transient intestinal upset.
	Half-life	Not applicable

Formulation: Vaporizer
Route of administration: Inhalational

Description of vaporizer: Inhaling herbs via a humidifier is one way to deliver herbs to the lungs and airways. As part of an asthma herbal treatment, a humidifier is a cheap and useful tool.

The way to use a humidifier is simply to add essential oils or tinctures to the water. Using reverse osmosis or distilled water is ideal. Glass and stainless steel are better materials than plastic when looking for a humidifier.

To conserve essential oils, sit directly above the humidifier to inhale the steam, and turn it off when you are not using it.

Eucalyptus: Eucalyptus is a tree. The dried leaves and oil are used to make medicine.

Eucalyptus	Dosage	The appropriate dose of eucalyptus depends on several factors such as the user's age, health, and several other conditions. At this time, there is not enough scientific information to determine an appropriate range of doses for eucalyptus.

	Use	Help to dilate the airways and also have an antibacterial effect. Eucalyptus **leaf** contains chemicals that might help control blood sugar and have activity against bacteria and fungi. Eucalyptus **oil** contains chemicals that might help pain and inflammation. It might also block chemicals that cause asthma.
	Patient group	Adults and children
	Side effects	Eucalyptus is safe in amounts found in foods. Eucalyptol, a chemical in eucalyptus oil, appears to be safe when taken by mouth for up to 12 weeks. Eucalyptus oil can cause nausea, vomiting, and diarrhea. Signs of eucalyptus poisoning might include stomach pain and burning, dizziness, muscle weakness, small eye pupils, feelings of suffocation, and some others.
	Half-life	Not applicable. Excreted in the lungs

Formulation: Intradermal (ID) injection
Route of administration: Injection

Description: The introduction of a hypodermic needle into the dermis for the purpose of instilling a substance, such as a serum or vaccine.

Tuberculin Skin Testing (Mantoux Test)

Tuberculin Skin Testing (Mantoux Test)	Dosage	A standard dose of five Tuberculin units (0.1 mL) is injected intradermally (between the layers of dermis) and read 48–72 hours later. A person who has been exposed to the bacteria is expected to mount an immune response in the skin containing the bacterial proteins. The reaction is read by measuring the diameter of induration (palpable raised hardened area) across the forearm (perpendicular to the long axis) in millimeters. If there is no induration, the result should be recorded as "0 mm." Erythema (redness) should not be measured.

Use	Tuberculin skin testing is a diagnostic tool for tuberculosis. It is one of the two major tuberculin skin tests used in the world, largely replacing multiple-puncture tests such as the Tine test.
Patient group	Adults and children
Side effects	Not applicable
Half-life	Not applicable. Not absorbed in Skin testing

Formulation: Intramuscular (IM) injection

Route of administration: Injection

Description of IM injection: It is the injection of a substance directly into a muscle. It is one of several alternative methods for the administration of medications. It is used for particular forms of medication that are administered in small amounts. Depending on the chemical properties of the drug, the medication may either be absorbed fairly quickly or more gradually. They should be given with extreme care, especially in the thigh, because the sciatic nerve may be injured or a large blood vessel may be entered if the injection is made without drawing back on the syringe first.

Bicillin C-R Tubex: Bicillin C-R (penicillin G benzathine and penicillin G procaine injectable suspension) contains equal amounts of the benzathine and procaine salts of penicillin G. It is available for deep intramuscular injection.

| Bicillin C-R Tubex | How supplied | Bicillin® C-R (penicillin G benzathine and penicillin G procaine injectable suspension) is supplied in packages of 10 TUBEX® Sterile Cartridge-Needle Units as follows: 1-mL size, containing 600,000 units per TUBEX® (21 gauge, thin-wall 1-inch needle for pediatric use), NDC 61570-139-10. 2-mL size, containing 1,200,000 units per TUBEX® (21 gauge, thin-wall 1-inch needle for pediatric use), NDC 61570-141-10. 2-mL size, containing 1,200,000 units per TUBEX® (21 gauge, thin-wall 1-1/4 inch needle), NDC 61570-140-10. |
| | Dosage | Bicillin® C-R (penicillin G benzathine and penicillin G procaine injectable suspension) is supplied in packages of 10 TUBEX® |

		Sterile Cartridge-Needle Units as follows: 1-mL size, containing 600,000 units per TUBEX® (21 gauge, thin-wall 1-inch needle for pediatric use)
		2-mL size, containing 1,200,000 units per TUBEX® (21 gauge, thin-wall 1-inch needle for pediatric use) 2-mL size, containing 1,200,000 units per TUBEX® (21 gauge, thin-wall 1-1/4 inch needle) Also available in packages of 10 disposable syringes as follow: 4-mL size, containing 2,400,000 units per syringe (18 gauge × 2-inch needle) Store in a refrigerator, 2°C–8°C (36°F–46°F). Keep from freezing.
	Use	Bicillin C-R is indicated in the treatment of the following in adults'and pediatric patients: Moderately severe to severe infections of the upper-respiratory tract, scarlet

		fever, erysipelas, skin and soft-tissue infections due to susceptible streptococci, moderately severe pneumonia and otitis media due to susceptible pneumococci. This antibiotic treats only bacterial infections. It will not work for viral infections (e.g., common cold, flu).
	Patient group	Adults and pediatric patients
	Side effects	Vision changes, fast/slow/pounding heartbeat, numbness/tingling of arms/legs, pain/redness/swelling of arms/legs, change in skin color near injection site or on arms/legs, uncontrolled movements, inability to move, change in the amount of urine, new signs of infection (e.g., fever, persistent sore throat), easy bruising/bleeding, extreme tiredness, dark/cloudy urine, seizures, mental/mood changes (e.g., depression, agitation).

		Seek immediate medical attention if any of these rare but very serious side effects occur: trouble breathing, chest pain, slurred speech, confusion, fainting.
	Half-life	½–1 hour

Formulation: Intraosseous (IR) injection
Route of administration: Injection

Description: Intraosseous infusion is the process of injection directly into the marrow of the bone. The needle is injected through the bone's hard cortex and into the soft marrow interior. Often the antero-medial aspect of the tibia is used because it lies just under the skin and can easily be palpated and located.

Examples

	Dosage	Bicillin® C-R (penicillin G benzathine and penicillin G procaine injectable suspension) is supplied in packages of 10 TUBEX® Sterile Cartridge-Needle Units as follows: 1-mL size, containing 600,000 units per TUBEX® (21 gauge, thin-wall 1-inch needle for pediatric use) 2-mL size, containing 1,200,000 units per TUBEX® (21 gauge, thin-wall 1-inch needle for pediatric use)

		2-mL size, containing 1,200,000 units per TUBEX® (21 gauge, thin-wall 1-1/4 inch needle) Also available in packages of 10 disposable syringes as follows: 4-mL size, containing 2,400,000 units per syringe (18 gauge × 2-inch needle) Store in a refrigerator, 2°C–8°C (36°F–46°F). Keep from freezing.
	Use	Bicillin C-R is indicated in the treatment of the following in adults and pediatric patients: Moderately severe to severe infections of the upper-respiratory tract, scarlet fever, erysipelas, skin and soft-tissue infections due to susceptible streptococci, moderately severe pneumonia and otitis media due to susceptible pneumococci. This antibiotic treats only bacterial infections. It will not work for viral infections (e.g., common cold, flu).
	Patient group	Adults and pediatric patients

	Side effects	Vision changes, fast/slow/pounding heartbeat, numbness/tingling of arms/legs, pain/redness/swelling of arms/legs, change in skin color near injection site or on arms/legs, uncontrolled movements, inability to move, change in the amount of urine, new signs of infection (e.g., fever, persistent sore throat), easy bruising/bleeding, extreme tiredness, dark/cloudy urine, seizures, mental/mood changes (e.g., depression, agitation). Seek immediate medical attention if any of these rare but very serious side effects occur: trouble breathing, chest pain, slurred speech, confusion, fainting.
	Half-life	½-1 hour

Formulation: Intraperitoneal (IP) injection

Route of administration: Injection

Description: IP injection is the injection of a substance into the peritoneum (body cavity). IP injection is more often applied to animals than humans. It is generally preferred when large amounts of blood-replacement fluids are needed, or when low blood pressure or other problems prevent the use of a suitable blood vessel for intravenous injection. It is faster than subcutaneous or intramuscular injection and used when veins are not accessible.

In humans, the method is widely used to administer chemotherapy drugs to treat some cancers, particularly ovarian cancer. This specific use has been recommended, controversially, as a standard of care. The needle is introduced into the upper flank and the syringe plunger withdrawn to ensure that the intestine has not been penetrated. The injected solution should run freely.

Chemotherapy Drugs to Treat Some Cancers

Chemotherapy Drugs to Treat Some Cancers		
	Dosage	1 mg/mL, 2 mg/mL, or 4 mg/mL
	Use	Treatment of disease by chemicals especially by killing microorganisms or cancerous cells
	Patient group	Children and adults whom have cancer
	Side effects	Myelosuppression (decreased production of blood cells), mucositis (inflammation of the lining of the digestive tract), and alopecia (hair loss)
	Half-life	39.9–70.6 hours

Formulation: Intravenous (IV) injection
Route of administration: Injection

Description: Intravenous therapy or IV therapy is the giving of liquid substances directly into a vein. The intravenous route is the fastest way to deliver fluids and medications throughout the body. Some medications, as well as blood transfusions and lethal injections, can only be given intravenously.

Loxilan

Loxilan	Dosage	300 mgI/mL 350 mgI/mL
	Use	Loxilan is a diagnostic contrast agent. It is injected intravenously before taking X-ray images to increase arterial contrast in the final image.
	Patient group	
	Side effects	Body: allergic reaction, asthenia, chest and back pain, edema of the neck, facial edema, pain, peripheral edema

| | | Cardiovascular: atrial fibrillation, syncope, tachycardia, vasodilation, ventricular extra systole
Digestive: anorexia, constipation, dyspepsia, dysphagia, Gl hemorrhage, ileus, liver failure
Nervous: hypotonia, nystagmus, paresthesia, somnolence, vertigo
Respiratory: dyspnea, pharyngitis, rhinitis
Skin: pruritus, sweating
Special senses: amblyopia, conjunctivitis, taste perversion, vision abnormality
Urogenital: anuria, dysuria, hematuria, infection of urinary tract, impairment of urination, kidney failure |
| | Half-life | 102 +/− 16.9 minutes |

Formulation: Subcutaneous or hypodermic injection

Route of administration: Injection

Description: Injection made into the subcutaneous tissues—the layer of skin directly below the dermis and epidermis, collectively referred to as the cutis. Although usually fluid medications are injected, occasionally solid materials, such as steroid hormones, are administered subcutaneously in small, slowly absorbed pellets to prolong their effect. Subcutaneous injections may be given wherever there is subcutaneous tissue, usually in the loose skin on the side of the chest or in the flank. Subcutaneous injections are highly effective in administering vaccines and such medications as insulin, morphine, diacetylmorphine, or goserelin.

Insulin

Insulin		
	Dosage	
	Use	Insulin is always necessary for type 1 diabetes because the body has no internal source of insulin. People with type 2 diabetes may also need insulin, particularly those who have difficulty controlling their diabetes with oral medications.
	Patient group	Children and adults
	Side effects	Taking too little or too much of the drug can result to hyperglycemia (loss of appetite, thirst, flushing, drowsiness, and a fruity odor on the breath) or hypoglycemia (dizziness, sweating, tremor, confusion, and hunger)
	Half-life	2.3+/−1.3 hours

Formulation: Ear drops
Route of administration: Otic

Description: Ear drops are a form of medicine used to treat or prevent ear infections, especially infections of the outer ear and ear canal.

Polysporin Eye and Ear Drops

Polysporin	Dosage	Apply one or two drops to the affected eye or ear, four times a day or more frequently as directed by your doctor.
	Use	Polymyxin/gramicidin contains a combination of antibiotics used for the treatment of certain types of infections caused by bacteria. The eye/ear drops are used to treat and prevent some types of external infections of the eye and ear.
	Patient group	
	Side effects	Signs of severe allergic reaction such as severe rash or hives; difficulty breathing; or swelling of the mouth, lips, tongue, or throat
	Half-life	5–6 hours

Formulation: Eye drops

Route of administration: Ophthalmic administration

Description: Eye drops are saline-containing drops used as a vector to administer medication in the eye. Eye drops sometimes do not have medications in them and are only lubricating and tear-replacing solutions.

Artificial Tears
Ophthacare

Artificial Tears	Dosage	
	Use	Dryness and irritation associated with deficient tear production in keratoconjunctivitis sicca (dry eyes). They are also used to moisten contact lenses and in eye examinations.
	Patient group	Generally adults
	Side effects	Eye pain Irritation Redness Vision changes Discomfort
	Half-life	5–6 hours

Ophthacare	Dosage	One to two drops, four to five times daily.
	Use	Ophthacare eye drops provide a cool and soothing effect and helps against eye irritations and strains. It can also be useful in managing infective, inflammatory, and allergic eye disorders.
		Ophthacare is effective in the management of infective and inflammatory eye disorders. Ophthacare also relieves congestion, and by virtue of its cooling effect, is beneficial in eyestrain. Ideal for contact lens wearers to use at bedtime. Ophthacare relieves the strains caused by contact lenses. The herbs, used in Ophthacare eye drops, are reported to have various pharmacological activities, which in combination has produced a synergistic effect in terms of antimicrobial and anti-inflammatory activities. Hence, Ophthacare eye drops are beneficial in patients with acute and chronic conjunctivitis, which appears as epidemic in certain seasons of the year.

Patient group	All ages
Side effects	Ophthacare is not known to have any side effects if taken as per the prescribed dosage.
Half-life	5–6 hours

Formulation: Nasal spray
Route of administration: Nasal

Description: Nasal sprays, or nasal mists, are used for the nasal delivery of a drug or drugs, either locally to generally alleviate cold or allergy symptoms such as nasal congestion or systemically.

Although delivery methods vary, most nasal sprays function by instilling a fine mist into the nostril by action of a hand-operated pump mechanism. The three main types available for local effect are: antihistamines, corticosteroids, and topical decongestants.

	Dosage	Nasal Spray by the intranasal route only
	Use	Management of the nasal symptoms of perennial nonallergic rhinitis in adult and pediatric patients aged 4 years and older
	Patient group	Generally adults
	Side effects	Common side effects may include minor nosebleed, burning or itching in your nose, sores or white patches.
	Half-life	Not applicable

Formulation: Mouth spray
Route of administration: Oral

Description:

Zolpimist (Zolpidem tartrate)

Zolpimist	Dosage	The recommended dose for adults is 10 mg once daily immediately before bedtime. The total Zolpimist dose should not exceed 10 mg per day.
		Elderly or debilitated patients may be especially sensitive to the effects of zolpidem tartrate. Patients with hepatic insufficiency do not clear the drug as rapidly as normal subjects. The recommended dose of Zolpimist in both of these patient populations is 5 mg once daily immediately before bedtime.
		Dosage adjustment may be necessary when Zolpimist is combined with other CNS-depressant drugs because of the potentially additive effects.

	Use	Zolpimist (zolpidem tartrate) Oral Spray is indicated for the short-term treatment of insomnia characterized by difficulties with sleep initiation. Zolpidem tartrate has been shown to decrease sleep latency for up to 35 days in controlled clinical studies. The clinical trials performed in support of efficacy were 4–5 weeks in duration with the final formal assessments of sleep latency performed at the end of treatment.
	Patient group	Generally adults
	Side effects	• Serious anaphylactic and anaphylactoid reactions • Abnormal thinking, behavior changes, and complex behaviors • Withdrawal • CNS-depressant effects
	Half-life	2.8 hours

	Important note	After taking Zolpimist, you may get up out of bed while not being fully awake and do an activity that you do not know you are doing. The next morning, you may not remember that you did anything during the night. You have a higher chance for doing these activities if you drink alcohol or take other medicines that make you sleepy with Zolpimist. Reported activities include: • driving a car ("sleep-driving") • making and eating food • talking on the phone • having sex • sleep-walking

Formulation: Douches
Route of administration:

Description:

	Dosage	Usual dosage is one insert (100,000 units nystatin) daily for two weeks.
	Use	An antimycotic polyene antibiotic obtained from Streptomyces noursei. Nystatin Vaginal Inserts, USP are effective for the local treatment of vulvovaginal candidiasis (moniliasis).
	Patient group	Adults
	Side effects	Nystatin is virtually nontoxic and nonsensitizing and is well tolerated by all age groups, even on prolonged administration. Rarely, irritation or sensitization may occur.
	Half-life	Generally 24 hours

Formulation: Cream

Route of administration: Dermal administration

Description: Emulsion of oil and water in approximately equal proportions. Penetrates stratum corneum outer layer of skin well.

Amcinonide Cream USP, 0.1%
Amcinonide Ointment USP, 0.1%

Amcinonide Cream	Dosage	Topical corticosteroids are generally applied to the affected area as a thin film from two to three times daily depending on the severity of the condition.
	Use	Topical corticosteroids are indicated for the relief of the inflammatory and pruritic manifestations of corticosteroid-responsive dermatoses. Used as anti-inflammatory and antipruritic agents.
	Patient group	Adults

	Side effects	Burning Itching Irritation Dryness Folliculitis Hypertrichosis Acneiform eruptions Hypopigmentation Perioral dermatitis Allergic contact dermatitis Maceration of the skin Secondary infection Skin atrophy Striae Miliaria
	Half-life	Not applicable. Absorbed by the skin

Formulation: Gel
Route of administration: Dermal administration

Description: Liquefies upon contact with the skin

Diclofenac Gel

Diclofenac Gel	Dosage	
	Use	Diclofenac Gel is used for treating pain in certain joints (e.g., in the knees or hands) caused by osteoarthritis. Diclofenac Gel is an NSAID. It may work by blocking certain substances in the body that are linked to inflammation. NSAIDs treat the symptoms of pain and inflammation. They do not treat the disease that causes those symptoms.
	Patient group	Adults and most age groups
	Side effects	Mild irritation at the application site Allergic reactions (rash, hives, or itching)
	Half-life	Generally 60 hours

Formulation: Liniment

Route of administration: Dermal administration

Description: Liniment is a medicated topical preparation for application to the skin. Preparations of this type are also called balms. Liniments are of a similar viscosity to lotions (being significantly less viscous than an ointment or cream) but unlike a lotion a liniment is applied with friction, a liniment is always rubbed in.

Liniments are typically sold to relieve pain and stiffness, such as from sore muscles or from arthritis. These liniments typically are formulated from alcohol, acetone, or similar quickly evaporating solvents, and contain counterirritant aromatic chemical compounds, such as methyl salicilate, benzoin resin, or capsaicin.

Examples

	Dosage	Capsaicin Topical (topical cream, topical lotion, topical kit, topical liquid, topical film, compounding powder, topical stick)
	Use	Temporary relief of muscle or joint pain caused by strains, sprains, arthritis, bruising, and backaches.
	Patient group	All Age groups
	Side effects	Common in all forms: Burning, itching, dryness, pain, redness, swelling, or soreness at the application site
	Half-life	Not applicable. Absorbed by the skin

Formulation: Ointment
Route of administration: Dermal administration

Description: Combines oil (80%) and water (20%). Effective barrier against moisture loss.

Examples

	Dosage	Generally 80% oil, 20% water
	Use	Therapeutic balms that are applied topically for various complaints; an ointment consists of a finely ground herbal powder in any of a number of oil bases, including almond oil, beeswax, lanolin, lard, petroleum jelly, and sesame oil.
	Patient group	All age groups
	Side effects	Dermatitis or skin inflammation. You may have an itchy rash, redness, or swelling. You may also have bumps or blisters that crust over or ooze clear fluid.
	Half-life	Not applicable. Absorbed by the skin

Formulation: Lotion

Route of administration: Dermal administration

Description: Oil-in-water emulsions using a substance such as Cetearyl alcohol to keep the emulsion together, but water-in-oil lotions are also formulated.

Calamine

Dosage	Generally 80% oil, 20% water
Use	Sunburn, eczema, rashes, poison ivy, chickenpox, insect bites, stings, also used as a mild antiseptic to prevent infections and an astringent to dry weeping or oozing blisters and acne abscesses.
Patient group	Generally > 2 years
Side effects	Very common (10% or more): Peeling, application site erythema. Common (1% to 10%): Dryness, pruritus, contact sensitization reactions. Uncommon (0.1% to 1%): Burning sensation.
Half-life	Not applicable. Absorbed by the skin

Formulation: Paste
Route of administration: Dermal administration

Description: Combines three agents: oil, water, and powder—an ointment in which a powder is suspended.

Examples

	Dosage	Applies to the following strengths: 10 mg (6%; 7.5%; 20%) and 6 mg (10%; 15%; 18.9%; 51%). Also applies to the following strengths with camphor and menthol: 3 mg, 4 mg, and 15 mg.
	Use	Pruritus, burns - external
	Patient group	Generally adults and children > 2
	Side effects	Mouth/gum irritation infrequently occurs. Most people using this medication do not have serious side effects.
	Half-life	31.07 +/−10.64 hours

Formulation: Skin patch

Route of administration: Dermal administration (transdermal)

Description: A transdermal patch or skin patch is a medicated adhesive patch that is placed on the skin to deliver a specific dose of medication through the skin and into the bloodstream.

Nicotine Patch
Pain Relief Patch

Nicotinell	Dosage	Nicotinell patch program: **Step 1** Nicotinell large = 21 mg/24 hours for 3–4 weeks then; **Step 2** Nicotinell medium = 14 mg/24 hours for 3–4 weeks then; **Step 3** Nicotinell small = 7 mg/24 hours for 3–4 weeks
	Use	Nicotinell patch is a 24-hour nicotine replacement medication. It releases nicotine to help with cessation of tobacco smoking. Nicotinell patches are used to help you stop smoking as part of an overall treatment program. When you apply the patch to your skin, nicotine

		passes through your skin and into your body. The Nicotinell patch program involves up to three steps, each with a different patch size. As your body adjusts to not smoking, you should gradually reduce the dose of Nicotinell over a period of up to a maximum of 12 weeks until you no longer need to use the patches.
	Patient group	Generally adults
	Side effects	Headache or dizziness Muscle discomfort Mild stomach upset Difficulty in sleeping
	Half-life	Nicotine ranges from 1 to 2 hours and cotinine's between 15 and 20 hours.

	Dosage	Transdermal patch
Pain Relief Patch	Use	An analgesic for severe pain. Pain Relief Patch delivers a constant dose of Glucosamine and Chondroitin over a 24-hour period and gives you no need to worry about taking pills throughout the day. Chondroitin Sulphate can reduce the activity of elastase, an enzyme released by white blood cells in inflamed joints, which breaks down elastic fibers and thus reduces the resiliency of cartilage. Preliminary findings indicate that chondroitin may increase joint mobility and slow cartilage loss.
		Glucosamine has relieved the symptoms of osteoarthritis more than any other one nutrient. Glucosamine is manufactured by the body and is primarily to help form the cushioning components of joint fluids and surrounding tissues. It thickens synovial fluid, making it more elastic;

		repairs the cartilage in damaged arthritic joints; and creates more support for joints, including the vertebrae. Besides helping to form the cartilage, tendons, ligaments, and synovial fluid in the joint, it also plays a role in the formation of nails, skin, eyes, bones, and heart valves. And, finally, it's involved in the mucous secretions of the digestive, respiratory, and urinary tracts.
	Patient group	Generally adults
	Side effects	There appear to be no adverse effects associated with taking glucosamine or chondroitin supplements.
	Half-life	Elimination HL after patch removal: 13–22 hours

Formulation: Rectal suppository
Route of administration: Suppository

Description: A rectal suppository is a drug delivery system that is inserted into the rectum where it dissolves. The suppository is inserted as a solid, and will dissolve inside the body to deliver the medicine

Examples

Dosage	80 mg, 120 mg, 325 mg, and 650 mg
Use	Used to deliver both systemically acting and locally acting medications.
Patient group	Generally adults
Side effects	Rectal irritation/burning, abdominal discomfort/cramps, or small amounts of mucus in the stool may occur.
Half-life	Mean elimination HL: 5 hours

Formulation: Vaginal suppository
Route of administration: Suppository

Description: A vaginal suppository is a drug delivery system that is inserted into vagina where it dissolves.

The suppository is inserted as a solid, and will dissolve inside the body to deliver the medicine; used to treat gynecological ailments.

Cleocin® Vaginal Ovules
Mycelex-G (Clotrimazole Vaginal)

Cleocin® Vaginal Ovules	How supplied	Carton of three suppositories with one applicator.
	Dosage	The recommended dose is one Cleocin Vaginal Ovule (containing clindamycin phosphate equivalent to 100 mg clindamycin per 2.5 g suppository) intravaginally per day, preferably at bedtime, for 3 consecutive days.
	Use	Cleocin Vaginal Ovules are indicated for 3-day treatment of bacterial vaginosis in non-pregnant women.
	Patient group	Non-pregnant women

	Side effects	**Urogenital system:** Vulvovaginal disorder, vaginal pain, and vaginal moniliasis. **Body as a whole:** Fungal infection. **Urogenital system**: Menstrual disorder, dysuria, pyelonephritis, vaginal discharge, and vaginitis/vaginal infection. **Body as a whole**: Abdominal cramps, fever, flank pain, generalized pain, headache, localized edema, and moniliasis. **Digestive system**: Diarrhea, nausea, and vomiting. **Skin**: rash, application-site pain, and application-site pruritis.
	Half-life	Elimination HL after dosing with the suppository was 11 hours. Range from 4 to 35 hours.

Mycelex-G (Clotrimazole Vaginal)	Dosage	Generally 100 mg per 2.5 g
	Use	Mycelex-G (clotrimazole vaginal) is an antifungal medication. It prevents fungus from growing. Mycelex-G is used to treat vaginal candida (yeast) infections.
	Patient group	Women
	Side effects	Allergic reaction (shortness of breath; closing of your throat; swelling of your lips, face, or tongue; or hives). Other, less-serious side effects may be more likely to occur. These include burning, itching, irritation of the skin, and an increased need to urinate.
	Half-life	Range from 4 to 35 hours

Formulation: Urethral suppository
Route of administration: Suppository

Description: A urethral suppository is a drug delivery system that is inserted into the urethra where it dissolves.

Alprostadil

	Dosage	125 mcg, 250 mcg, 500 mcg, and 1000 mcg
	Use	Treatment of severe erectile dysfunction
	Patient group	Generally adults
Alprostadil	Side effects	Aching in the penis, testicles, legs, and in the perineum (area between the penis and rectum). Warmth or burning sensation in the urethra. Redness of the penis due to increased blood flow. Minor urethral bleeding or spotting due to improper administration.
	Half-life	Less than 10 minutes

Formulation: Liquid suppository
Route of administration: Suppository

Description: The activity of injecting a
liquid with a small syringe into the
rectum

Laxative

	Dosage	5.6 g rectally once and 2 to 3 g rectally once
Laxative	Use	Treat constipation
	Patient group	Generally adults
	Side effects	Upset stomach, stomach cramps, gas, diarrhea, burning, and rectal irritation
	Half-life	Mean elimination HL: 1.78 hours

Factors that may affect the choice of dosage forms:

1. Therapeutic Group of the Drug or Device
2. Patient Anatomy, Age, and Body Size
3. Presenting Disease or Illness
4. Recommended Route of Administration
5. Dosage Schedule of the Drug/Device
6. Documented Safety Profile of the Drug or Device
7. Documented Adverse/Side Effect Profile of the Drug/Device

Bibliography

Prozac

Cipriani, A., et al. "Comparative Efficacy and Acceptability of 21 Antidepressant Drugs for the Acute Treatment of Adults with Major Depressive Disorder: A Systematic Review and Network Meta-Analysis." *The Lancet*, vol. 391, no. 10128, 2018, pp. 1357–66. doi:10.1016/S0140-6736(17)32802-7.

Prozac FDA Product Monograph. 2011, https://www.accessdata.fda.gov/drugsatfda_docs/label/2011/018936s091lbl.pdf.

OCPs

Planned Parenthood. *How Effective Is the Birth Control Pill?* https://www.planned-parenthood.org/learn/birth-control/birth-control-pill/how-effective-is-the-birth-control-pill. Accessed May 19, 2019.

Rivera, R., et al. "The Mechanism of Action of Hormonal Contraceptives and Intrauterine Contraceptive Devices." *American Journal of Obstetrics and Gynecology*, vol. 181, no. 5 Pt 1, 1999, pp. 1263–69. doi:10.1016/s0002-9378(99)70120-1.

Benadryl

Benadryl. *BENADRYL® Allergy ULTRATAB® Tablets.* https://www.benadryl.com/products/benadryl-allergy-ultratab-tablets. Accessed May 19, 2019.

Benadryl. *BENADRYL® DOSING GUIDE.* https://www.benadryl.ca/prevention-tips/dosing-guide. Accessed May 19, 2019.

Benadryl. *BENADRYL® Caplets.* https://www.benadryl.ca/products/benadryl-caplets. Accessed May 19, 2019.

DrugBank. *Diphenhydramine.* https://www.drugbank.ca/drugs/DB01075. Accessed May 19, 2019.

Codeine

Codeine FDA Product Monograph. 2009, https://www.accessdata.fda.gov/drugsatfda_docs/label/2009/022402s000lbl.pdf.

Drendel, A.L., and S. Ali. "Efficacy and Practicality of Codeine." *CMAJ*, vol. 183, no. 3, 2011, p. 349. doi:10.1503/cmaj.111-2012.

Doxycycline

Doxycycline FDA Product Monograph. 2008, https://www.accessdata.fda.gov/drugsatfda_docs/label/2008/050795s005lbl.pdf.

Lithium

Lithium Carbonate FDA Product Monograph. 2018, https://www.accessdata.fda.gov/drugsatfda_docs/label/2018/017812s033,018421s032,018558s027lbl.pdf.

Sani, G., et al. "Treatment of Bipolar Disorder in a Lifetime Perspective: Is Lithium Still the Best Choice?" *Clinical Drug Investigation*, vol. 37, no. 8, 2017, pp. 713–27. doi:10.1007/s40261-017-0531-2.

Listerine

Listerine. *LISTERINE POCKETPAKS® Oral Care Breath Strips.* https://www.listerine.com/products/listerine-go/listerine-pocketpaks-oral-care-strips. Accessed May 30, 2019.

Cordran

Allergan. *Cordran® Tape Product Monograph*. 2018, https://www.allergan.com/assets/pdf/cordran_pi.

Goutos, I., and R. Ogawa. "Steroid Tape: A Promising Adjunct to Scar Management." *Scars, Burns & Healing*, vol. 3, 2017, p. 2059513117690937. doi:10.1177/2059513117690937.

Celestone

Betamethasone (Systemic). https://www.glowm.com/resources/glowm/cd/pages/drugs/b012.html. Accessed May 30, 2019.

CELESTONE® FDA Product Monograph. 2006, https://www.accessdata.fda.gov/drugsatfda_docs/label/2006/014215s009s015lbl.pdf.

Clemastine

RxList. *Clemastine Fumarate Syrup Product Monograph*. 2017, https://www.rxlist.com/clemastine-fumarate-syrup-drug.htm.

Schran, H.F., et al. "The Pharmacokinetics and Bioavailability of Clemastine and Phenyl-propanolamine in Single-Component and

Combination Formulations." *Journal of Clinical Pharmacology*, vol. 36, no. 10, 1996, pp. 911–22.

Turner, R.B., et al. "Effectiveness of Clemastine Fumarate for Treatment of Rhinorrhea and Sneezing Associated with the Common Cold." *Clinical Infectious Diseases: An Official Publication of the Infectious Diseases Society of America*, vol. 25, no. 4, 1997, pp. 824–30. doi:10.1086/515546.

Midazolam

RxList. *Midazolam Hydrochloride Syrup Product Monograph.* 2019, https://www.rxlist.com/midazolam-hydrochloride-syrup-drug.htm.

Salem, K., et al. "Efficacy and Safety of Orally Administered Intravenous Midazolam Versus a Commercially Prepared Syrup." *Iranian Journal of Pediatrics*, vol. 25, no. 3, 2015, p. e494. doi:10.5812/ijp.25(3)2015.494.

Rosemary

Habtemariam, S. "The Therapeutic Potential of Rosemary (Rosmarinus Officinalis) Diterpenes for Alzheimer's Disease." *Evidence-Based Complementary and Alternative Medicine: ECAM*, vol. 2016, 2016, p. 2680409. doi:10.1155/2016/2680409.

Nordqvist, J. "Everything You Need to Know about Rosemary." *MedicalNewsToday*, 2017, https://www.medicalnewstoday.com/articles/266370.php.

Ozarowski, M., et al. "*Rosmarinus officinalis* L. Leaf Extract Improves Memory Impairment and Affects Acetylcholinesterase and Butyrylcholinesterase Activities in Rat Brain." *Fitoterapia*, vol. 91, 2013, pp. 261–71. doi:10.1016/j.fitote.2013.09.012.

Flovent

FLOVENT HFA FDA Product Monograph. 2010, https://www.accessdata.fda.gov/drugsatfda_docs/label/2010/021433s015lbl.pdf.

Advair

ADVAIR DISKUS FDA Product Monograph. 2017, https://www.accessdata.fda.gov/drugsatfda_docs/label/2017/021077s056s057lbl.pdf.

Budesonide

PLUMICORT FLEXHALER FDA Product Monograph. 2010, https://www.accessdata.fda.gov/drugsatfda_docs/label/2010/021949s006lbl.pdf.

O'Connell, E.J. "Efficacy of Budesonide in Moderate to Severe Asthma." *Clinical Therapeutics*, vol. 24, no. 6, 2002, pp. 887–905; discussion 837.

Aerospan

AEROSPAN FDA Product Monograph. 2017, https://www.accessdata.fda.gov/drugsatfda_docs/label/2017/021247s015lbl.pdf.

VA Pharmacy Benefits Management Services, et al. *Flunisolide Inhalation Aerosol 80 mcg (Aerospan) Abbreviated Review.* 2014, https://www.pbm.va.gov/PBM/clinicalguidance/abbreviatedreviews/Flunisolide_Aerospan.docx.

AccuNeb

AccuNeb® (Albuterol Sulfate) Inhalation Solution FDA Product Monograph. 2011, https://www.accessdata.fda.gov/drugsatfda_docs/label/2011/020949s024lbl.pdf.

Intal

Intal® Nebulizer Solution FDA Product Monograph. 2004, https://www.accessdata.fda.gov/drugsatfda_docs/label/2004/18596slr030_intal_lbl.pdf.

"Cromoglicic Acid." *Wikipedia, the Free Encyclopedia*. https://en.wikipedia.org/wiki/Cromoglicic_acid. Accessed June 6, 2019.

Nebupent

NebuPent FDA Product Monograph. 2011, https://www.accessdata.fda.gov/drugsatfda_docs/label/2011/019887s014lbl.pdf.

Drugs.com. *Pentamidine (Oral Inhalation)*. 2019, https://www.drugs.com/ppa/pentamidine-oral-inhalation.html.

Lobelia

Bell, R.L., et al. "Nicotinic Receptor Ligands Reduce Ethanol Intake by High Alcohol-Drinking HAD-2 Rats." *Alcohol (Fayetteville, N.Y.)*, vol. 43, no. 8, 2009, pp. 581–92. doi:10.1016/j.alcohol.2009.09.027.

Drugs.com. *Lobelia*. 2019, https://www.drugs.com/npp/lobelia.html.

Icahn School of Medicine at Mount Sinai. *Lobelia*. https://www.mountsinai.org/health-library/herb/lobelia. Accessed June 13, 2019.

LLC, WebMD. *Lobelia*. https://www.webmd.com/vitamins/ai/ingredientmono-231/lobelia. Accessed June 13, 2019.

Stansbury, J., et al. "The Use of Lobelia in the Treatment of Asthma and Respiratory Illness." *Journal of Restorative Medicine*, vol. 2, no. 1, 2013, pp. 94–100.

Cayenne pepper

"Capsaicin." *ScienceDirect*. https://www.science-direct.com/topics/neuroscience/capsaicin. Accessed June 13, 2019.

Viokase

de la Iglesia-García, D., et al. "Efficacy of Pancreatic Enzyme Replacement Therapy in Chronic Pancreatitis: Systematic Review and Meta-Analysis." *Gut*, vol. 66, no. 8, 2017, pp. 1354LP–1355. doi:10.1136/gutjnl-2016-312529.

DrugBank. *Pancrelipase*. https://www.drugbank.ca/drugs/DB00085. Accessed June 13, 2019.

Prescribers' Digital Reference. *Pancrelipase—Drug Summary*. https://www.pdr.net/drug-summary/Viokace-pancrelipase-1522. Accessed June 13, 2019.

RxList. *Viokase Product Monograph*. 2008, https://www.rxlist.com/viokase-drug.htm.

Eucalyptus

DrugBank. *Eucalyptol*. https://www.drugbank.ca/drugs/DB03852. Accessed June 13, 2019.

Tyagi, A.K., and A. Malik. "Liquid and Vapour-Phase Antifungal Activities of Selected Essential Oils against Candida Albicans: Microscopic Observations and Chemical Characterization of Cymbopogon Citratus." *BMC Complementary and Alternative Medicine*, vol. 10, 2010, p. 65. doi:10.1186/1472-6882-10-65.

Wong, C. "The Health Benefits of Eucalyptus Oil." *VerywellHealth*, 2019, https://www.verywellhealth.com/steam-inhalation-with-eucalyptus-essential-oil-88169.

Tubersol

Tuberculin Purified Protein Derivative (Mantoux) FDA Product Monograph. https://www.fda.gov/media/74866/download. Accessed June 13, 2019.

DrugBank. *Tuberculin Purified Protein Derivative.* https://www.drugbank.ca/drugs/DB11601. Accessed June 13, 2019.

RxList. *Tubersol Product Monograph.* 2018, https://www.rxlist.com/tubersol-drug.htm.

Bicilin C-R

DrugBank. *Procaine Benzylpenicillin.* https://www.drugbank.ca/drugs/DB09320. Accessed June 20, 2019.

RxList. *Bicilin C-R Tubex Product Monograph.* 2019, https://www.rxlist.com/bicillin-c-r-tubex-drug.htm.

Ioxilan

Oxilan FDA Product Label. 2017, https://www.accessdata.fda.gov/drugsatfda_docs/label/2017/020316s029lbl.pdf.

Insulin

"Insulin (Medication)." *Wikipedia, the Free Encyclopedia.* https://en.wikipedia.org/wiki/Insulin_(medication). Accessed June 20, 2019.

Church, T.J., and S.T. Haines. "Treatment Approach to Patients with Severe Insulin Resistance." *Clinical Diabetes: A Publication of the American Diabetes Association*, vol. 34, no. 2, 2016, pp. 97–104. doi:10.2337/diaclin.34.2.97.

Drugs.com. *Insulin Regular Dosage.* 2019, https://www.drugs.com/dosage/insulin-regular.html.

Duckworth, W.C., et al. "Insulin Degradation: Progress and Potential*." *Endocrine Reviews*, vol. 19, no. 5, 1998, pp. 608–24. doi:10.1210/edrv.19.5.0349.

Erpeldinger, S., et al. "Efficacy and Safety of Insulin in Type 2 Diabetes: Meta-Analysis of Randomised Controlled Trials." *BMC Endocrine Disorders*, vol. 16, no. 1, 2016, p. 39. doi:10.1186/s12902-016-0120-z.

Healthline. *Long-Acting Insulin: How It Works*. https://www.healthline.com/health/diabetes/long-acting-insulin. Accessed June 20, 2019.

Polysporin

Kwa, A.L.H., et al. "Pharmacokinetics of Polymyxin B in a Patient with Renal Insufficiency: A Case Report." *Clinical Infectious Diseases: An Official Publication of the Infectious Diseases Society of America*, vol. 52, no. 10, 2011, pp. 1280–81. doi:10.1093/cid/cir137.

Polysporin. *POLYSPORIN Pain Relief Ear Drops*. https://www.polysporin.ca/products/pain-relief-ear-drops. Accessed June 20, 2019.

RxList. *Xylocaine Product Monograph*. 2018, https://www.rxlist.com/xylocaine-drug.htm.

Artificial Tears

Drugs.com. *Artificial Tears*. 2018, https://www.drugs.com/mtm/artificial-tears.html.

Pucker, A.D., et al. "Over the Counter (OTC) Artificial Tear Drops for Dry Eye Syndrome." *The Cochrane Database of Systematic Reviews*, vol. 2, 2016, p. CD009729. doi:10.1002/14651858.CD009729.pub2.

Ophthacare

Biswas, N.R., et al. "Evaluation of Ophthacare Eye Drops—A Herbal Formulation in the Management of Various Ophthalmic Disorders." *Phytotherapy Research: PTR*, vol. 15, no. 7, 2001, pp. 618–20.

Zolpimist

Zolpimist FDA Product Monograph. 2008, https://www.accessdata.fda.gov/drugsatfda_docs/label/2008/022196lbl.pdf.

DrugBank. *Zolpidem*. https://www.drugbank.ca/drugs/DB00425. Accessed June 22, 2019.

Neubauer, D.N. "ZolpiMist™: A New Formulation of Zolpidem Tartrate for the Short-Term Treatment of Insomnia in the US." *Nature and Science of Sleep*, vol. 2, 2010, pp. 79–84.

Amcinonide

Amcinonide. https://www.drugbank.ca/drugs/DB00288. Accessed June 22, 2019.

Bickers, D.R. "A Comparative Study of Amcinonide and Halcinonide in the Treatment of Eczematous Dermatitis." *Cutis*, vol. 34, no. 2, 1984, pp. 190–94.

Cornell, R.C. "Comparison of Amcinonide Ointment 0.1 Percent Twice Daily and Fluocinonide Ointment 0.05 Percent Three Times Daily in the Treatment of Psoriasis." *Cutis*, vol. 31, no. 5, 1983, pp. 566–69.

Engel, M.F. "Treatment of Psoriasis with Amcinonide 0.1 Percent and Fluocinonide 0.05 Percent Ointments. A Comparative Double-Blind Study." *Cutis*, vol. 29, no. 6, 1982, pp. 646–50.

Guenther, L., et al. "A Controlled Comparison of Amcinonide Cream 0.1 Percent and Halcinonide Cream 0.1 Percent in the Treatment of Eczematous Dermatitis." *Cutis*, vol. 28, no. 4, 1981, pp. 461–462, 464, 467.

Limited, Teva Canada. *Ratio-Amcinonide Product Monograph*. 2017, https://pdf.hres.ca/dpd_pm/00039966.PDF.

Rosenberg, E.W. "Management of Eczematous Dermatitis with Amcinonide or Betamethasone Valerate. A Double-Blind Comparative Study." *Cutis*, vol. 24, no. 6, 1979, pp. 642–45.

Diclofenac

VOLTAREN GEL FDA Product Monograph. 2016, https://www.accessdata.fda.gov/drugsatfda_docs/label/2016/022122s010lbl.pdf.

Pharmacy Times. *Should Topical NSAIDs Have Strict Heart Risk Warnings?* 2015, https://www.pharmacytimes.com/contributor/jeffrey-fudin/2015/07/should-topical-nsaids-have-strict-heart-risk-warnings.

Zeng, C., et al. "Relative Efficacy and Safety of Topical Non-Steroidal Anti-Inflammatory Drugs for Osteoarthritis: A Systematic Review and Network Meta-Analysis of Randomised Controlled Trials and Observational Studies." *British Journal of Sports Medicine*, vol. 52, no. 10, 2018, pp. 642LP–650. doi:10.1136/bjsports-2017-098043.

Calamine

"LABEL: CALAMINE—Ferric Oxide Red Lotion." *DailyMed, U.S. National Library of Medicine.* 2011, https://dailymed.nlm.nih.gov/dailymed/drugInfo.cfm?setid=8c1b0675-bbe9-42a4-80f1-3efff1d6346c.

"Calamine." *Wikipedia, the Free Encyclopedia.* https://en.wikipedia.org/wiki/Calamine. Accessed June 22, 2019.

DrugBank. *Ferric Oxide.* https://www.drugbank.ca/drugs/DB11576. Accessed June 22, 2019.

Medicine India. *Calamine Pharmacology.* https://www.medicineindia.org/pharmacology-for-generic/1872/calamine. Accessed June 22, 2019.

Tebruegge, M., et al. "Does the Use of Calamine or Antihistamine Provide Symptomatic Relief from Pruritus in Children with Varicella Zoster Infection?" *Archives of Disease in Childhood*, vol. 91, no. 12, 2006, pp. 1035–36. doi:10.1136/adc.2006.105114.

Nicotinell

Fiore, M.C., et al. "The Effectiveness of the Nicotine Patch for Smoking Cessation. A Meta-Analysis." *JAMA*, vol. 271, no. 24, 1994, pp. 1940–47.

Gupta, S.K., et al. "Bioavailability and Absorption Kinetics of Nicotine Following Application of a Transdermal System." *British Journal of Clinical Pharmacology*, vol. 36, no. 3, 1993, pp. 221–27. doi:10.1111/j.1365-2125.1993.tb04221.x.

Nicotinell. *Nicotinell Nicotine Patch*. http://nicotinell.ie/products/nicotinell-patches/. Accessed June 22, 2019.

Walker, N., et al. "Effectiveness and Safety of Nicotine Patches Combined with E-Cigarettes (with and without Nicotine) for Smoking Cessation: Study Protocol for a Randomised Controlled Trial." *BMJ Open*, vol. 9, no. 2, 2019, p. e023659. doi:10.1136/bmjopen-2018-023659.

Pain Relief Patch

DrugBank. *Chondroitin Sulfate*. https://www.drug-bank.ca/drugs/DB09301. Accessed June 22, 2019.

Kirkham, S.G., and R.K. Samarasinghe. "Review Article: Glucosamine." *Journal of Orthopaedic Surgery (Hong Kong)*, vol. 17, no. 1, 2009, pp. 72–76. doi:10.1177/230949900901700116.

PatchMD. *Glucosamine & Chondroitin Topical Patch (30-Day Supply)*. https://www.patchmd.com/Glucosamine-Chondroitin-Topical-Patch.html. Accessed June 22, 2019.

PatchMD. *PatchMD FAQ's*. https://www.patchmd.com/Patch-Education.html. Accessed June 22, 2019.

Persiani, S., et al. "Glucosamine Oral Bioavailability and Plasma Pharmacokinetics after Increasing Doses of Crystalline Glucosamine Sulfate in Man." *Osteoarthritis and Cartilage*, vol. 13, no. 12, 2005, pp. 1041–49. doi:10.1016/j.joca.2005.07.009.

Shmerling, R.H. "The Latest on Glucosamine/Chondroitin Supplements." *Harvard Health Blog*. 2016, https://www.health.harvard.edu/blog/the-latest-on-glucosaminechon-droitin-supplements-2016101710391.

Cleocin

NDA 050767/S-010 Cleocin® Vaginal Ovules. 2012.

Institute for Quality and Efficiency in Health Care, (IQWiG). "Which Treatments Are Effective for Bacterial Vaginosis?" *InformedHealth. Org [Internet]*, 2009.

Broumas, A.G., and L.A. Basara. "Potential Patient Preference for 3-Day Treatment of Bacterial Vaginosis: Responses to New Suppository Form of Clindamycin." *Advances in Therapy*, vol. 17, no. 3, 2000, pp. 159–66.

Mycelex G

3 Day Vaginal Cream; Clotrimazole Vaginal Cream USP (2%); Product Label. 2016, https://www.accessdata.fda.gov/drugsatfda_docs/label/2016/021143Orig1s012lbl.pdf.

DrugBank. *Clotrimazole*. https://www.drugbank.ca/drugs/DB00257. Accessed July 2, 2019.

LLC, WebMD. *Mycelex-G Tablet*. https://www.webmd.com/drugs/2/drug-7443/mycelex-g-vaginal/details. Accessed July 2, 2019.

O-Prasertsawat, P., and A. Bourlert. "Comparative Study of Fluconazole and Clotrimazole for the Treatment of Vulvovaginal Candidiasis." *Sexually Transmitted Diseases*, vol. 22, no. 4, 1995, pp. 228–30.

Zhou, X., et al. "The Efficacy and Safety of Clotrimazole Vaginal Tablet vs. Oral Fluconazole in Treating Severe Vulvovaginal Candidiasis." *Mycoses*, vol. 59, no. 7, 2016, pp. 419–28. doi:10.1111/myc.12485.

Alprostadil

"LABEL: MUSE-Alprostadil Suppository." *DailyMed, U.S. National Library of Medicine*, https://dailymed.nlm.nih.gov/dailymed/drugInfo.cfm?setid=4c55f3f9-c4cf-11df-851a-0800200c9a66. Accessed July 2, 2019.

DrugBank. *Alprostadil*. https://www.drugbank.ca/drugs/DB00770. Accessed July 2, 2019.

LLC, WebMD. *How Does Alprostadil Treat Erectile Dysfunction?* https://www.webmd.com/erectile-dysfunction/guide/alprostadil-treat-ed#1. Accessed July 2, 2019.

Laxative

"LABEL: FLEET-Glycerin Suppository." *DailyMed, U.S. National Library of Medicine*, https://dailymed.nlm.nih.gov/dailymed/drugInfo.cfm?setid=669fa039-d683-49f3-aeeb-d687ccd5f74e. Accessed July 2, 2019.

LLC, WebMD. *Safely Using Laxatives for Constipation*. https://www.webmd.com/digestive-disorders/laxatives-for-constipation-using-them-safely#1. Accessed July 2, 2019.

Index

A

active birth control pills, 7–8
Advair Diskus inhaler, 54–60
Aerobid inhaler, 62–65
aerosol, 50
    Flovent (fluticasone propionate), 50–53
Albuterol Sulfate inhalation, 66–68
alprostadil, 126
amcinonide cream/ointment, 110–111
antimicrobials, 72
artificial tears, 101

B

balms, 113, 115
Benadryl, 14–15
betamethasone, 35–41
Bicillin C-R Tubex, 86–90
birth control pills, 7–12
blow molding, 19
botanical medicine, 49
breakthrough bleeding, 9
budesonide, 61

C

D

E

F

G

H

S

T

U

V

Z